Wine Production and Quality

Wine Production and Quality

Keith Grainger and Hazel Tattersall

SECOND EDITION

WILEY Blackwell

Registered Office
John Wiley & Sons, Ltd, The Atrium, Southern Gate, Chichester, West Sussex, PO19 8SQ, UK

Editorial Offices
9600 Garsington Road, Oxford, OX4 2DQ, UK
The Atrium, Southern Gate, Chichester, West Sussex, PO19 8SQ, UK
111 River Street, Hoboken, NJ 07030-5774, USA

For details of our global editorial offices, for customer services and for information about
how to apply for permission to reuse the copyright material in this book please see our
website at www.wiley.com/wiley-blackwell.

Library of Congress Cataloging-in-Publication Data

Names: Grainger, Keith, author. | Tattersall, Hazel, author.
Title: Wine production and quality / Keith Grainger, Hazel Tattersall.
Other titles: Wine production
Description: Second edition. | Chichester, West Sussex, UK ; Hoboken, NJ : John Wiley & Sons
  Inc., [2016] | Revised edition of: Wine production / Keith Grainger and Hazel Tattersall. Oxford ;
  Ames, Iowa : Blackwell Pub., 2005. | Includes bibliographical references and index.
Identifiers: LCCN 2015037524 | ISBN 9781118934555 (cloth)
Subjects: LCSH: Wine and wine making.
Classification: LCC TP548 .G683 2016 | DDC 663/.2–dc23
LC record available at http://lccn.loc.gov/2015037524

A catalogue record for this book is available from the British Library.

Contents

Preface

This book comprises a single volume on the topics of wine production, tasting and quality. In 2005, *Wine Production: Vine to Bottle* was published by Blackwell. This book became required reading for many basic oenology and more general wine courses held at institutions in several countries around the world. A Spanish-language edition, *Producción de Vino*, was published by Editorial Acribia. The year 2009 saw the publication by Wiley of *Wine Quality: Tasting and Selection*; this work won the Gourmand Award for Best Wine Education Book in the World for that year, and subsequently the Gourmand Award for the Best Wine Education Book in the years 1995–2014. *Wine Production and Quality* is a combined second edition of these books. In the years since the publication of the previous works, there has been much change in the wine industry, and the perceptions of critics and expectations of consumers have also shifted. Accordingly, the text has been revised and expanded, and there is much new material. The book is divided into two parts: Part 1 – Wine Production and Part 2 – Wine Quality.

There are, of course, many detailed books on the topics of grape growing and winemaking, and the multifarious individual aspects thereof. These books, although very valuable to oenology students, grape growers and winemakers, are often highly scientific or technical. There are also several 'coffee table' books that paint a picture that may be appreciated by consumers and those with merely a passing interest. In Part 1 of this book, 'Wine Production', we aim to provide a concise, structured yet readable understanding of wine production, together with a basic source of reference. Although the content includes necessary scientific information, it is designed to be easily understood by those with little scientific knowledge.

In Part 2 of the book, 'Wine Quality', we aim to provide an understanding of the concepts and techniques of tasting, assessing and evaluating wines for their styles and qualities, and of the challenges in assessing and recognising quality in wines. We also discuss the faults that can destroy wines at any quality level and the misconceptions as to what constitutes quality. As with Part 1, the text is written primarily for the reader with limited scientific knowledge, but at times it is necessary to take a more scientific approach, especially when examining the compounds that give rise to aromas, flavours and, particularly, taints. The text is also unashamedly interspersed with the occasional anecdote, for it is not just our personal perceptions but also our experiences that shape our interaction with what can be the most exciting of beverages.

The tasting structure and tasting terms used are generally those of the *Systematic Approach to Tasting of the Diploma Level* of the Wine & Spirit Education Trust. Accordingly, we hope the book may prove valuable to those studying for, or considering studying for, this internationally recognised qualification.

The book does not examine grape varieties in detail, or other than by way of example, the profiles and qualities of the vast array of wines produced in the wine regions of the world. There is already a wealth of literature on these topics. We briefly look at Champagne, Sherry and Port as the leading examples of wines made by their particular methods. However, there are many references to Bordeaux and its wines. Our reasons are straightforward: Bordeaux is the largest 'fine' wine region in the world. Its reputation has been largely built upon the unsurpassed excellence of the Grands Crus Classés wines, although the majority of the production is of 'everyday Bordeaux'. The region remains a benchmark, flagship and model to winemakers and wine lovers around the world.

The information contained in this book is not from any parochial or polarised viewpoint. However, the authors, like all wine lovers, cannot (and do not wish to) claim to be totally objective. During the research and preparation, we have spent much time in wine regions in both the Old and New Worlds. We listened to diverse and detailed viewpoints from many hundreds of practitioners, including growers and vineyard workers in both cool and hot climates, family winery owners, winemakers and technicians working with large-scale producers, consultants and representatives of wine institutes. Accordingly, we believe *Wine Production and Quality* will prove valuable to food and beverage industry professionals, wine-trade students, wine merchants, sommeliers, restaurateurs and wine lovers, as well as those entering (or thinking of entering) the highly competitive world of wine production.

We wish to thank everybody who has given their time, knowledge and opinions. We also wish especially to thank Antony Moss MW and Trevor Elliott for reviewing the text and making valuable suggestions. We also wish in particular to thank the Wine & Spirit Education Trust for allowing us to use, adapt and extract from the *WSET® Systematic Approach to Tasting – Diploma*.

Keith Grainger
Hazel Tattersall

Acknowledgements

Figures 1.2, 1.4, 1.5, 4.2: Christopher Willsmore
Figure 4.10: Zuccardi Argentina
Figure 7.1: Château Lassegue
Figure 15.1: Champagne Taittinger/Hatch Mansfield
Figure 15.2: Brett Jones
Figure 16.2: Riedel
Figure 25.5: AMOS INDUSTRIE
Figure 25.10: Zuccardi Argentina
All other figures: Keith Grainger

Acknowledgement is given to Wine & Spirit Education Trust for allowing the use of the *WSET® Systematic Approach to Tasting – Diploma*.

PART 1

Introduction to Part 1 – Wine Production

No other beverage is discussed, adored or criticised in the same way as wine. To a few, it is something to be selected with the greatest of care, laid down until optimum maturity, carefully prepared for serving, ritually tasted in the company of like-minded people using a structured technique and then analysed in the manner of both the forensic scientist and literary critic. To many, it is simply the bottle bought in the supermarket according to the offer of the moment, drunk and perhaps enjoyed on the same day as purchased. To those favoured with living in wine-producing regions, it is often the beverage purchased from the local producers' cooperative from a dispenser resembling a petrol pump, taken home in a 5- or 10-litre container and drunk with each and every meal.

There is a wonderful diversity in the styles and quality of wines produced throughout the world, promoting discussion and disagreement among wine lovers. The wines of individual producers, regions and countries rise and fall in popularity according to consumer, press and TV media perceptions of style, quality, fashion and value. Consumers do not remain loyal when they perceive that their needs and wants are better met elsewhere. If we consider the United Kingdom wine market, back in the 1980s, red wines from Bulgaria were very popular, and white German wines held the No. 1 position in the league table for white wine sales by volume. Australian wines were almost unheard of. By 2005, the wines of Australia held the top position in the UK wine market, by both volume and value of sales. In 2015, Australia still led the field in UK, although, in a market that suffered some decline over the previous 10 years, the volume of sales had slipped by 17%.

Few would dispute that the standard of wines made today is higher than at any time in the 8000 years or so of vinous history. The level of knowledge

Wine Production and Quality, Second Edition. Keith Grainger and Hazel Tattersall.
© 2016 John Wiley & Sons, Ltd. Published 2016 by John Wiley & Sons, Ltd.

of producers, and thus the ability to control the processes in wine production, could only have been dreamt of even 40 years ago. Yet when, a few years ago, *Decanter* magazine compiled a list of the greatest wines of all time, the top position was awarded to Château Mouton-Rothschild 1945, and six of the 'top 10' wines were produced more than 40 years ago. Also, in the past few years, globalisation and consolidation of producers have perhaps had the detrimental effect of producing technically good wines whose styles have become standardised. In other words, the wonderful diversity we referred to is under threat.

In this part of the book, we detail how wine is produced, from vine to bottle. Many of the concepts are simple to grasp, others more complex. However, we need to stress at this stage that there is no single, unquestioned approach to wine production. Many procedures in common usage remain subject to challenge. Indeed, if you talk to 50 winemakers, you are likely to hear 100 different viewpoints, and many producers are constantly experimenting and changing techniques.

In considering wine production, there are two distinct stages: the growing of grapes (viticulture) and turning grapes into wine (vinification). Throughout the wine-producing world, there are many in the industry who carry out just one of these stages. There are growers who make no wine but sell their grapes to a wine-producing firm, or who are members of a cooperative that will make the wine. There are also wine producers who have no vineyards, or insufficient vineyards to supply their grape needs and consequently buy grapes from growers small or large. The decisions made and operations undertaken in both the vineyard and winery will affect the style and quality of the finished wine. These decisions will be based on numerous factors: geographical, geological, historical, legal, financial and commercial. The resources and availability and cost of local labour will have a major impact upon the decisions made and the structure of the wine-production operation. Both the grower and the winemaker are aiming for maximum control: yield, quality, style and cost. Of course, the aim is to make a profit.

Grapes contain all that is basically necessary to make wine: the pulp is rich in sugar, and yeasts are present in the bloom on the skins. These yeasts also migrate onto winery surfaces and may initiate a spontaneous fermentation of the sugar rich *must*. Must may be defined as grape juice and solids prior to fermentation. However, many winemakers choose to inhibit these natural yeasts and use cultured yeasts for fermentations. It should be noted that, unlike in the production of beer (and many spirits), water is not generally used as an ingredient in winemaking. The grapes should be freshly gathered, and ideally the winemaking should take place in the district of origin. However, this is not always adhered to, particularly with regard to inexpensive wines. It is not uncommon for grapes or grape must to travel from one region to another, or sometimes even to another country, prior to fermentation.

Wine is, of course, alcoholic. The alcohol in wine is ethanol, otherwise known as ethyl alcohol. Although it is a natural product, ethanol is toxic and can damage the body if taken in excess. The alcohol is obtained from the fermentation of must by the action of enzymes of yeast that convert the grape sugars into ethanol and carbon dioxide. Although the fermentation lies at the heart of winemaking, every other operation will impact upon the finished wine. The entire production process may take as little as a few weeks for inexpensive wines, or two years or more for some of the highest-quality wines. In the case of some fortified wines, the production process may take over a decade.

Throughout this book we will usually refer to the area measurement of land in hectares, which is the most used term in member states of the European Union, although the United Kingdom prefers to measure in acres. A hectare is 2.47 acres, some 10,000 square metres of land. Units of liquid measure are expressed in litres (l) and hectolitres (hl) – there are 100 litres in a hectolitre. Units of weight will be stated in grams (g), kilograms (kg) and metric tonnes, i.e. 1000 kg.

The methods and techniques explained in this part of the book include some that have been recently introduced by forward-thinking producers. There is a constant strive for improvement at all levels in the industry. As South African winemaker Beyers Truter says: 'If you ask me: "Have you made the best wine, or the best wine that you can?" and I answer "Yes", then you must take me away and bury me.'

CHAPTER 1

Viticulture – the basics

The aim of the grape grower is, following a successful annual vineyard cycle, to harvest ripe and healthy grapes, of the quality and to the specification required for subsequent vinification. The grower and winemaker are both aware that any deficiencies in the quality of the fruit will affect not only the quality of the wine but also profitability. In this chapter, we will examine the grape vine and its fruit in some detail. We also look at the reasons why vines are grafted onto rootstocks, including the devastating effect of the *Phylloxera* louse, and why crossings have been developed.

1.1 The grape vine

The cultivation of the grape vine is known to have begun some 8000 years ago in the Near East. Archaeological evidence of cultivated grape pips has been found in the Republic of Georgia and dated 6000–7000 BC. A potsherd (fragment of pottery) found in Iran and dated around 5000 BC has been analysed and found to contain salt from tartaric acid, which could only have come from grape juice, and resin used as a wine preservative. Wine presses from 2000 to 3000 BC have been found in south-eastern Turkey. In the ensuing millennia, viticulture spread throughout Europe and parts of Asia, and, in the last 230–460 years, also to New World countries.

The grape vine belongs to a family of climbing flowering plants called Vitaceae (formerly Ampelidaceae). The family comprises 15 genera, including the genus *Vitis*, the grape-bearing vine. This genus comprises some 65 species, including *Vitis vinifera*. It is worth noting that the members of any species have the ability to exchange genes and to interbreed. *V. vinifera* is the European and

Wine Production and Quality, Second Edition. Keith Grainger and Hazel Tattersall.
© 2016 John Wiley & Sons, Ltd. Published 2016 by John Wiley & Sons, Ltd.

central Asian species of grapevine, and it is from this species that almost all of the world's wine is made.

1.2 Grape varieties

V. vinifera has, as we now believe, some 10,000 different varieties, e.g. *V. vinifera* Chardonnay, *V. vinifera* Cabernet Sauvignon. Each variety looks different and tastes different. Some varieties ripen early, others late; some are suitable for growing in warm climates, others prefer cooler conditions; some like certain types of soil, others don't; some yield well, others are extremely shy bearing. Some can produce first-class wine, others distinctly mediocre. An illustration of some of the grape varieties planted in Argentina is shown in Figure 1.1.

Whilst these are all factors of relevance to a grower, the actual choice of variety or varieties planted in any vineyard may well, as in the European Union (EU), be determined by wine laws. For example, red Beaune must be made from the variety Pinot Noir. It is worth remembering that most of the varieties that we know have been cultivated and refined by generations of growers, although some such as Riesling are probably the descendants of wild vines.

The grape variety, or blend of grape varieties, from which a wine is made is a key factor in determining the design, style, aromas and flavours of the wine. Wines made from a single variety are sometimes referred to as varietals. The name of the variety may be stated on the label, this concept having been introduced in Alsace in the early 1920s and promoted heavily by the Californian producers in the 1970s, and has now become commonplace. However, many

Figure 1.1 Some grape varieties planted in Argentina.

wines made from a single variety do not state the fact on the front label, e.g. a bottle of Chablis will rarely inform you that the wine is made from Chardonnay. Many top-quality wines are made from a blend of two or more varieties, with each variety helping to make a harmonious and complex blend. This may perhaps be compared with cooking, where every ingredient adds to taste and balance. Examples of well-known wines made from a blend of varieties include most red Bordeaux, which are usually made from two to five different varieties (Cabernet Sauvignon, Cabernet Franc, Merlot, Malbec, Petit Verdot), and red Châteauneuf-du-Pape where up to 13 can be used.

Of the 10,000 or so different grape varieties, only 500 or so are commonly used for winemaking. The names of just a few of these, e.g. Sauvignon Blanc, are very well known. Some varieties are truly international, such as Chardonnay, which is planted in many parts of the world. Others are found in just one country, or even one region within a country such as the Mencia variety in north-west Spain. Many varieties have different names in different countries and even pseudonyms in different regions of the same country. So, for example, southern Portugal's Fernão Pires changes its name to Maria Gomes further north in Bairrada, and Croatia's Trbljan has perhaps 13 synonyms within the country.

Discussion of the characteristics of individual grape varieties is a detailed topic and is beyond the scope of this book. For further information, the reader is referred to the Bibliography.

1.3 The structure of the grape berry

Although the juice of the grape is seen as the essential ingredient in the winemaking process, other grape constituents also have roles of varying importance, and we will briefly examine these, including their impact upon the wine produced.

Figure 1.2 shows a section through a typical ripe grape berry.

1.3.1 Stalks

A cluster of grape berries includes a considerable amount of stems (stalks). The individual stalk of each berry is the pedicel, which is attached to the rachis, or main axis stem of the cluster. The cluster is attached to the vine by the peduncle, and it is this stem that is usually cut by the grape picker, if the fruit is being harvested by hand. Stalks contain tannins that may give a bitter taste and an astringent feel to wine. Whether or not the stems are included in early stages of the winemaking process is a matter of choice, depending on the style required. The winemaker may choose to destem the grapes completely before they are crushed. Alternatively, the stalks, or just a small proportion of them, may be left on to increase the tannin in red wine to give

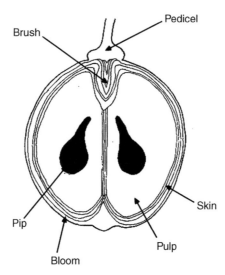

Figure 1.2 Structure of the grape berry. Source: Courtesy of Christopher Willsmore.

extra structure. Also, if the stalks are not removed, they perform a useful task in the pressing operation by acting as drainage channels, helping to prevent juice pockets.

1.3.2 Skins

Skins contain colouring matters, aroma compounds, flavour constituents and tannins. There are several layers of skin; the outside waxy layer with its whitish hue is called bloom. This contains yeasts and bacteria. Below this, we find further layers containing complex substances called polyphenols (a class of flavonoids), which can be divided into two groups:

1 *Anthocyanins (black grapes)* and *flavones (white grapes)* give grapes their colour, and as phenolic biflavanoid compounds, they form antioxidants which help preserve the wine and perhaps provide health-giving properties. The predominant anthocyanin in *V. vinifera* grapes is malvidin 3,5-diglucoside $(C_{29}H_{35}O_{17})$.

2 *Tannins* are bitter compounds that are also found in stalks and pips. They can, if unripe or incorrectly handled, give an excessively dry, green or bitter mouth feel on the palate. Tannin levels are higher in red wines where more use is made of the skins in the winemaking process and with much greater extraction from the skins than in white and rosé wines. Tannin gives full-bodied red wines 'grip' and firmness in the mouth. Some varieties such as Cabernet Sauvignon, Syrah and Nebbiolo contain high levels of tannins; others such as Gamay have much lower levels.

1.3.3 Yeasts

Yeasts are single-celled micro-organisms belonging to the Fungi kingdom. Their usual method of reproduction is by budding. There are over 1500 species, of which just a few are of interest to the winemaker. The enzymes in yeasts are, of course, essential for the wine-fermentation process. Yeasts, together with bacteria, attach themselves to the bloom on the skins of grapes. There are two basic groups of yeast present on the skins:

- Wild yeasts: these are mostly of the genera *Kloeckera* and *Hanseniaspora*. Wild yeasts only operate aerobically. Once in contact with grape sugars, they can convert these sugars to alcohol, but only up to approximately 4% alcohol by volume (abv), at which point they die.
- Wine yeasts, of the genus *Saccharomyces* (sugar fungus). These can operate both aerobically and anaerobically. During a fermentation, they may continue to work until either there is no more sugar left in the juice or an alcoholic strength of approximately 15% has been reached, at which point they die naturally.

1.3.4 Pulp

The pulp or flesh contains juice. If you peel the skin of either a green or black skinned grape, the colour of the flesh is not dissimilar. The actual juice of the grape is almost colourless, with the very rare exception of a very few varieties that have tinted flesh, e.g. Gamay Teinturier and Dunkelfelder. The pulp contains water, sugars, fruit acids, proteins and minerals:

- Water: approximately 70–80% of the grape pulp is water.
- Sugars: when unripe, all fruits contain a high concentration of acids and low levels of sugar. As the fruit ripens and reaches maturity, the balance changes, with sugar levels rising and acidity falling. Photosynthesis is the means by which a greater part of this change occurs. Grape sugars are mainly represented by fructose and glucose, with each comprising between 8 and 12% of the weight of a ripe berry. Sucrose, although present in the leaves and phloem tubes of the vine, has no significant presence in the berry because having been transported into the grape, it is hydrolysed into its constituents. As harvest nears, the producer can measure the rise in sugar levels by using a refractometer, as illustrated in Figure 1.3.
- Acids: tartaric acid and malic acid account for between 69 and 92% of the acidity of the grape berry, the latter being of a higher proportion in unripe grapes. During the ripening process, the amount of malic acid decreases, and tartaric becomes the principal acid. In fact, the amount of tartaric acid remains constant, but it is diluted as the grape berry swells. Tartaric acid is not found in significant quantities in any other cultivated fruits of European origin, although it is a component in bananas, mangos and tamarinds. Acids have an important role in giving wine a refreshing, mouth-watering taste and also give stability and longevity to the finished product. There are

Figure 1.3 Refractometer.

tiny amounts of other organic acids present in grapes, including acetic, citric and succinic acids. Amino acids are also present in tiny amounts, mainly arginine and proline.

- Minerals: potassium is the main mineral present in the grape pulp, with a concentration of up to 2500 mg/l. Of the other minerals present, none has a concentration of more than 200 mg/l, but the most significant are calcium, magnesium and sodium.

1.3.5 Pips

Pips or seeds vary in size and shape according to grape variety. Unlike with stalks, there is no means of separating them at reception at the winery and, if crushed, will impart astringency to the wine owing to their bitter oils and hard tannins. As we shall see later, modern presses are designed to minimise this happening. We will discuss grape-berry development in Chapter 4.

1.4 Crossings, hybrids, clonal and massal selection

1.4.1 Crossings

It is possible to cross two varieties of *V. vinifera* (by fertilising one variety with the pollen of another variety) and thus produce a *crossing*, itself a new variety. For example, the variety Marselan is a crossing of Cabernet Sauvignon and Grenache, and was bred in 1961 by the French National Institute for Agricultural Research (INRA). The aim of breeding the crossing was to create a disease-resistant variety with the heat tolerance of Grenache together with the elegance and finesse of Cabernet Sauvignon. It should be noted, however, that a crossing will not necessarily inherit the characteristics of its parent varieties.

1.4.2 Hybrids

It is important not to confuse the term *crossing* with *hybrid*. A hybrid is a 'crossing' of two vine species. Hybrids, in the first half of the twentieth century, were planted extensively in France, but during the 1960s and 1970s, some 325,000 hectares (ha) were grubbed up. The use of hybrids is prohibited in the EU for the production of wines of the Protected Designation of Origin (PDO) category, in theory the highest quality level.

1.4.3 Clones and massal selection

From any variety, breeders can select individual clones. Clonal selection is basically breeding asexually from a single parent, aiming to obtain certain characteristics such as yield, flavour, good plant shape, early ripening, disease resistance, etc. Each vine will be identical in DNA and 'personality'. Planting of vineyards with single clones of a variety became commonplace in the 1970s to 1990s. However, in spite of the tremendous development with clonal selection during the past 40 years, many growers believe that old vines give the highest-quality juice. Massal selection involves taking cuttings from outstanding and perhaps old vines in a vineyard and propagating the budwood. This process is now regaining popularity. Research on genetic modification of vines is taking place, but at present no wine is produced from genetically modified plants.

1.5 Grafting

Although nearly all the world's wine is produced by various varieties of the species *V. vinifera*, the roots on which the *V. vinifera* vines are growing are usually those of another species, or hybrids of other species. Why is this? *V. vinifera* has already been described as the European species of vine, because of its origin. Other species of grapevine exist whose origins are elsewhere, particularly in the Americas, e.g. *Vitis rupestris*. However, although these other species produce grapes, the wine made from them has a most unpleasant taste. Therefore, *V. vinifera* is the species that is used to produce almost all of the world's wine.

At the time that the countries in the New World were colonised, it was obvious that wine could be produced in parts of many of them, but *V. vinifera* needed to be brought from Europe in order to produce palatable wines. However, *V. vinifera* had no resistance to the many pests and diseases that existed in the New World. Disastrously, during the late nineteenth century, these pests and diseases found their way from the USA to Europe on botanical specimens and plant material. The vineyards of Europe suffered initially from mildews, which will be detailed in Chapter 5, and then from a most lethal pest, *Phylloxera vastatrix*, that eats into the roots of *V. vinifera*, resulting in the vines dying.

1.6 *Phylloxera vastatrix*

Of all the disasters that have struck the vineyards of Europe over the centuries, the coming of *Phylloxera vastatrix* (recently reclassified as *Daktulosphaira vitifoliae*; see Chapter 5) was by far the most devastating. It was first discovered in 1863 in a greenhouse in Hammersmith, London, and named *Phylloxera* in 1868, the same year as the pest was found in vineyards in the Rhône valley in France. It was noted in Bordeaux in 1869, and by 1877 it had arrived in Geelong in Victoria, Australia. It was found in New Zealand in 1885.

Phylloxera had lived in North America, east of the Rocky Mountains, for thousands of years. Naturally the American species of vines had become resistant, the roots having developed the ability to heal over after attack. As the pest wound its devastating way throughout the vineyards of France, into Spain and elsewhere during the latter years of the nineteenth century, many attempts were made to find a cure. Poisoning the soil and flooding the vineyards were two of the ideas tried, which now seem extreme. Many growers simply gave up, and in many regions the amount of land under vines shrank considerably. In the Chablis district of Burgundy, there were some 40,000 ha of vines prior to the arrival of *Phylloxera*, and owing to the invasion of the pest (together with crop losses owing to spring frosts) this had shrunk to 550 ha by the end of the Second World War.

Phylloxera is an aphid that has many different manifestations during its life cycle, and can breed both asexually and sexually. An illustration of *Phylloxera* in its root-living form is shown in Figure 1.4. When reproducing asexually, eggs are laid on the vine roots and hatch into crawlers. These develop into either wingless or winged adults. *Phylloxera* only feeds on the roots of *V. vinifera*. The vine's young, feeder roots develop nodosities – hook-shaped galls. The older, thicker, storage roots develop wart-like tuberosities. The cambium and phloem are destroyed, preventing sap from circulating and synthesised food being transmitted downward from the leaves. Root tips can be killed by a process of strangulation by the canker. It must be stressed that *Phylloxera* affects the very life of the vine rather than the quality of the grapes and resulting wine.

As we have seen, the grapes of American species of vines make unpleasant-tasting wine. However, if a *V. vinifera* plant is grafted onto an American species rootstock (or a hybrid of two American species), then the root is resistant to attack. The grafted plant yields quality grapes perfectly suitable for winemaking. The primary functions of the roots are to provide anchorage for the vine and to draw up water and nutrients, including trace elements that will help give the wine flavours. So, producers almost the world over began grafting their vines. The economic cost of *Phylloxera* is huge – the cost of purchasing grafted vines from a nursery may be four times that of ungrafted material.

Although the vast majority of the world's vines are now grafted, there are still several areas where ungrafted vines are common, including Chile, areas of Argentina and most of the Mosel in Germany. *Phylloxera* will not live in certain

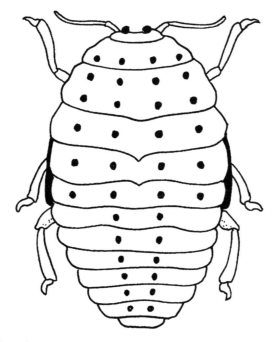

Figure 1.4 *Phylloxera* louse. Source: Courtesy of Christopher Willsmore.

soil types, including sand and slate. The states of Western Australia, South Australia and Tasmania are *Phylloxera*-free. For *Phylloxera* to spread, it is usually transmitted on plant materials, grapes, clothing, tools, picking bins or vehicles – unless transported, the aphid will spread only a hundred metres or so in a season. So, as in Chile and the *Phylloxera*-free parts of Australia (known as *Phylloxera* Exclusion Zones), a rigorous quarantine programme has excluded the louse, for its arrival in the vineyards could spell economic disaster. To prevent any risk of the aphid spreading on equipment, this must be thoroughly disinfected or heat treated. For example, tractors may be put into a chamber and heated to 45°C for 2 h; picking bins may be heated to 70°C for 5 min. New vine planting material must also be heat treated: 50°C for 5 min is effective.

1.7 Rootstocks

There are many different American species of vines, and just a few of these are suitable for rootstocks. Three of the most widely used are *Vitis berlandieri*, *Vitis riparia* and *Vitis rupestris*. The choice of species for rootstock will depend not only on the vine stock, but also on the climate and soil type. Most commercially used rootstocks are hybrids of two American species. Table 1.1 describes several rootstocks in common usage. In some regions, the grafting

Table 1.1 Some rootstocks in common usage

Rootstock name	Hybridisation	Characteristics
Richter 110	*berlandieri/rupestris*	Fairly vigorous. Suitable for shallow soils. Resistant to 17% active limestone. Moderately drought-tolerant.
Ruggeri 140	*berlandieri/rupestris*	Vigorous. Grows in exhausted soils. Resistant to 30% active limestone. Very drought-tolerant.
Paulsen 1103	*berlandieri/rupestris*	Very vigorous. Fair resistance to salinity. Resistant to 17% active limestone. Moderately drought-tolerant.
SO4	*berlandieri/riparia*	Average vigour. Suitable for poorly drained soils. Resistant to nematodes. Resistant to 20% active limestone.
101–14	*riparia/rupestris*	Average vigour. Suitable for alluvial soils. Low resistance to active lime.

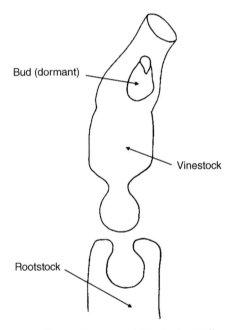

Figure 1.5 Graft – omega cut. Source: Courtesy of Christopher Willsmore.

of vines still takes place in the vineyards, but it is usual for producers to buy new vines from specialist nurseries, where they have been bench-grafted. The grower wishing to establish a new vineyard, or replace existing vines, will order from the nursery a specified clone of a variety of *V. vinifera* pre-grafted onto a suitable species or hybrid of root stock. The most common type of graft used nowadays is the machine-made omega cut, as illustrated in Figure 1.5.

Some rootstocks that were once considered to be *Phylloxera*-resistant (e.g. rootstock AXR1) were found not be, in the late twentieth century. Expensive lessons were learned by many a Californian grower. It is also argued that in recent decades, there has evolved a new, even more troublesome biotype of *Phylloxera vastatrix*. Before leaving the topic of rootstocks, we should note that in addition to the primary functions discussed above, roots produce hormones, including cytokinins and gibberellins that stimulate and regulate growth, and control other plant functions.

1.8 The life of the vine

Grapevines can live for over 100 years, but the lifespan will depend upon many factors, including the climate of the country/region in which they are planted, the soil type and the methods of viticulture. In the first few years after planting (3–5 years) as the vine establishes its root system, yields will be naturally low. Even so, such small yields from the young vine can sometimes give intense flavours, and classic varietal character.

As vines grow older, their root systems become more complex with deeper penetration of the ground in search of water and nutrients. It is generally accepted that older vines give particularly good fruit which, of course, can lead to more concentrated and intense wines. After about 20 years, vines start to become less vigorous, thus producing smaller yields. Accordingly, some producers have a replanting cycle to ensure there is no drop in production. They may, for example, decide to replant vineyards when the vines have reached a certain age – perhaps 30–40 years or so. In areas where there are problems with virus, a vineyard could be grubbed up when the vines are just 20 years old. In the best vineyards, producers will often replace diseased or dying vines on an individual basis, to retain a high average age. Ungrafted vines can have the longest life of all, and there are a few areas in the world where some vines aged 120 years or more still thrive, such as the Grosse Lage (Grand Cru) vineyards of the Dr. Loosen Estate, in the Mosel, Germany.

CHAPTER 2

Climate

Climate and weather have a major influence upon the quality and style of wines produced. This chapter considers the key factors for consideration in terms of suitable climates for viticulture, climatic challenges and the impact upon the styles of wine produced under different climatic conditions. We also look at the topic of climate change.

Most of the world's major wine-producing regions are to be found between 30° and 50° latitude in both northern and southern hemispheres. Within this 'temperate zone', grapes will usually ripen satisfactorily; however, there is a wide range of climatic variation. By way of a simple comparison, we might contrast the cool, damp and sometimes humid climatic conditions of northern Europe with those of the extreme south of Europe and North Africa, where grapes may struggle in the dryness and heat. The overall climate of any vineyard region may be referred to as the macroclimate. However, as we shall see, climate does vary within areas of any region, and even small variations have an important impact upon the quantity, style and quality of grapes produced.

2.1 World climate classifications

The Kőppen system of climate classification remains the most widely used method of classifying the world's climates, over 100 years after its inception. Under the system, climates are classified into five bands, each of which has sub-divisions. The five broad bands (with the sub-divisions of the bands of interest to us) are:

- A – tropical moist climates: all months have average temperatures above 18°C;
- B – dry climates: with deficient precipitation during most of the year; this band is divided into Bw (dry arid) and Bs (dry semi-arid);

Wine Production and Quality, Second Edition. Keith Grainger and Hazel Tattersall.
© 2016 John Wiley & Sons, Ltd. Published 2016 by John Wiley & Sons, Ltd.

- C – moist mid-latitude climates with mild winters; this band is divided into Cfa (humid subtropical), Cfb (martime) and Cs (Mediterranean);
- D – moist mid-latitude climates with cold winters;
- E – polar climates: with extremely cold winters and summers.

As we shall see below, it is generally areas within Bs, Cfa, Cfb and Cs that are suitable for viticulture.

2.2 Climatic requirements of the grape vine

2.2.1 Sunshine

Ideally, the vine needs a minimum of 1400 h of sunshine per year, with a minimum average of 6–7 h per day during the growing season (April to October in the northern hemisphere and October to April in the southern hemisphere). Regions with too much sunshine and heat tend to yield wines that are coarse with high levels of alcohol and low levels of acidity. Too little sunshine leads to unripe grapes, which in turn results in wines that are over high in acidity and light in alcohol, and with dilute and often green flavours. The vine produces sugars by the biological and biochemical process of photosynthesis, using light energy to synthesise water and carbon dioxide to produce carbohydrates. However, for growing ripe grapes, the vine needs sunshine more for its heat than for its light.

2.2.2 Warmth

Growth of the vine takes place only at temperatures above 10°C (50°F). For the vine to flower successfully in early summer, a temperature of 15°C (59°F) is needed. The period from flowering to harvesting averages 100 days, but could be as short as 80 days or as long as 150 days. During this period, minimum average temperatures of 18°C (64°F) are needed for white grapes to ripen and 20°C (68°F) for reds. Average daily summer temperatures of much above 23°C (73°F) can have a negative impact on grape flavours, depending upon the variety, although many fine wine regions exceed this in the height of summer, e.g. parts of Napa in California average 26.2°C (79.2°F), and parts of Pasa Robles, also in California, average 30.3°C (86.5°F).

2.2.3 Cold winter

The vine needs to rest in winter. Without this winter rest, the vine might yield twice a year, and its life would be shortened. Winter frost can be beneficial for hardening wood and may kill fungal diseases and insect pests. There are, however, vineyards in Brazil where five crops are obtained in a 2-year period – the climate is hot all year round, and the growing seasons are simulated by turning the irrigation on and off. The vines grown in this way quickly become exhausted and need replacement.

2.2.4 Rainfall

Obviously, vines need water to grow. An annual rainfall of between 500 mm and 850 mm is perhaps ideal depending upon, inter alia, soil types and other climatic conditions, otherwise irrigation will usually be necessary. Winter rain builds up water reserves underground, as the plants are in a dormant state. During the growing season, at least 300 mm of rain (or equivalent irrigation) is required. Rain in early spring helps the vine's growth; in summer and early autumn, it swells the fruit. In Europe, nature normally provides sufficient rainfall, but in New World countries, irrigation is commonplace. For example, the relatively new wine-producing region of Limarí in Chile is technically a desert, parts of which have less than 100 mm of rain a year. The area is arid (Kőppen Division Bw), but the vine flourishes by irrigation. However, there are semi-arid areas in New World countries where dry land wine farming is practised, but deep-rooted vines are considered essential. Just before harvest, a little rain can help increase the yield, but it needs to be followed by warm sunshine and a gentle drying breeze, if undesirable rots are to be prevented. Heavy rain will often split berries, causing severe damage and diluting juice.

2.3 Climatic enemies of the grape vine

2.3.1 Frost

Severe winter frosts can damage vines. Temperatures below −16°C (3°F) can kill the vine by freezing the sap, resulting in roots splitting. This happened in Bordeaux in February 1956, resulting in much of the vineyard area having to be replanted. Frost at budding time can destroy the young buds and shoots, sometimes considerably reducing the year's crop, as happened in 1991 in Bordeaux, where, as a result of a devastating frost on the night of 21 April, crop losses in Saint-Émilion and Pomerol were as much as 80%. On 30 April 2015, a severe frost hit many vineyards in Turkey, particularly the Manisa region, wiping out most of the year's harvest. Known areas for frost (frost pockets), particularly sites that have poor air circulation, are perhaps best avoided for new plantings, although some classic wine districts, e.g. parts of Chablis, can produce wines of such quality that the producers have reluctantly borne the risk.

There are several methods of frost protection, but all have a cost in labour and equipment. In winter, the base of the vine's trunk can be earthed up to prevent damage to the graft. Various measures can be used to mitigate the effects of spring frosts. Traditionally, oil burners were widely used in the vineyard at budding time to circulate air, but these are now considered primitive. Wind machines can prove effective. An illustration of a wind machine is shown in Figure 2.1. In New Zealand, which has had an increasing problem with frosts since 2001, helicopters are often used to circulate the air. It is not uncommon for there to be over 100 helicopters in the air over the region of

Figure 2.1 Wind machine in New Zealand.

Marlborough, which is situated in the north of New Zealand's South Island. Many producers choose to use an *aspersion system* to prevent damage – this involves installing a sprinkler system, which sprays water onto the buds before the air temperature falls below 0°C (32°F). The theory is simple: water contains heat, and when the water freezes, much of this heat will go into the bud, which will then be protected in its own igloo of an ice pellet. However, no frost-prevention system is totally effective.

2.3.2 Hail

Hail usually falls in relatively small, localised areas and can result in severe physical damage to both vines and fruit. This can range from scarring of leaves, bruising or breaking of young shoots (the effects of which can carry on to the following season), to the splitting of berries. Many classic regions may suffer hail damage. On 19 May 2014, hail hit the vineyard of Château Les Ormes De Pez in the Bordeaux district of Saint-Estèphe, destroying leaves and shoots and resulting in a reduction of 30% in the year's harvest. On 25 July 2014, hail hit further south in Bordeaux at Saint Saveur, resulting in substantial crop losses of Blanc de Lynch-Bages. If hail occurs closer to the harvest, split or smashed berries are susceptible to rot or may start to ferment on the vine, resulting in whole bunches being unusable. Various forms of protection

Figure 2.2 Trellised vines netted for hail protection.

include fine netting, either as an overhead canopy or as vertical nets against the trellis system. An illustration of trellised vines in Mendoza, Argentina, netted for hail protection is shown in Figure 2.2. More controversially, the firing of rockets carrying silver nitrate into clouds may cause the ice to fall as rain. For growers in affected areas, insurance is an expensive option.

2.3.3 Strong winds

These can have a dramatic effect by damaging canes, breaking shoots and removing leaves. Severe winds in spring, in particular, can be devastating for young shoots and leaves or very young vines. If prevalent when the vine is flowering, poor pollination and a reduced crop can result. Strong winds might lead to *coulure*, a condition where the flowers remain closed, and pollination fails. This leads to grape clusters having a reduced number of berries, as illustrated in Figure 2.3.

The detrimental effect of strong winds may be a particular problem in valleys, which can act as funnels. For example, on the steeply terraced hillsides

Figure 2.3 A sparse grape cluster resulting from Coulure.

in France's northern Rhône valley, vines have to be individually staked, while in the flatter southern Rhône, rows of conifer trees have been planted to break the destructive force of the Mistral wind. In other areas, forests and mountain ranges can offer some measure of protection.

2.3.4 Excessive heat

Heat stress can be harmful to the vine. When there is excessive sun, and temperatures exceed 40°C (104°F), the vine can shut down. Grapes can be sunburnt and scarred in the hot afternoon sun, so a grower may choose to leave more leaves on the western side of the vines. Excessive heat may result in wines with 'cooked' flavours, unbalanced and with high levels of alcohol.

2.3.5 Drought

Drought can also result in a reduced crop, as happened in Europe (especially much of France) in 2003. Even with established irrigation systems in place, New World countries can suffer, too. This was the case with the 2007 harvest in Australia, where the lack of rainfall caused restrictions in irrigation water availability, and in 2014 in parts of California, especially the Central Coast.

2.4 Mesoclimate and microclimate

There is often confusion between these two terms. Mesoclimate refers to the local climate within a particular vineyard or part of a vineyard. Microclimate refers to the climate within the canopy of leaves that surrounds the vine. There are several factors that influence the mesoclimate of any vineyard area.

2.4.1 Water

Proximity to water, whether rivers, seas, lakes, or reservoirs, can sometimes bring the vines the benefit of reflected heat. Water can act as a heat reservoir, releasing during the night the heat stored by day. This has the double advantage of moderating temperature and reducing risk of frost. Water also encourages mists and high humidity which can lead to mildews (always unwelcome) or rot (occasionally sought if the production of sweet white wines is the objective).

2.4.2 Altitude

For every 100 m of altitude, the mean air temperature decreases by approximately 0.6°C (1.1°F). Thus, a grower may choose to plant varieties that prefer cooler climates at higher altitudes, e.g. Sauvignon Blanc or Riesling. Higher altitudes can result in greater diurnal ranges, as discussed in Chapter 23.

2.4.3 Aspect

When the vineyard is situated on a slope, the direction, angle and height of the slope are important, and the slope may provide protection from prevailing winds. Frost at budding time is less likely to be a problem, since it tends to roll down slopes. Slopes can be beneficial in aiding ripening. It is perhaps worth reflecting on some basic botany and biochemistry: the vine sucks in carbon dioxide (CO_2) from the atmosphere and combines it with water (H_2O) from the soil to create carbohydrate ($C_6H_{12}O_6$) in the form of grape sugars. This process is known as assimilation – there is spare oxygen (O_2) left over that is transpired to the atmosphere. In the northern hemisphere, the further a vineyard is situated from the equator, the lower the sun is in the sky, and the greater the advantage of a south-east-facing slope. The vines are inclined towards the morning sun when CO_2 in the atmosphere is at its highest level, resulting in the manufacture of more sugars. An illustration of a steep sloped vineyard in the Mosel Valley, Germany, is shown in Figure 2.4.

Figure 2.4 Steep-sloped vineyard in the Mosel.

2.4.4 Woods and trees

Groups of trees can protect vines from strong winds but can have the unwelcome effect of encouraging high humidity. They also decrease the diurnal temperature range and can reduce the air temperature.

2.5 The concept of degree days

In 1944, a system of climatic classification based on degree days (the Winkler system) was devised by Amerine and Winkler of the Viticultural School of the University of California at Davis, as a result of their research in matching vines to climate. A calculation is based on the 7 months' annual growing season according to the hemisphere:

- northern hemisphere: theoretical growing season = 1 April to 31 October;
- southern hemisphere: theoretical growing season = 1 November to 30 April.

The vine does not grow at temperatures below 10°C (50°F). Each day in the growing season that the mean temperature is above 10°C is counted and totalled. Each degree of mean temperature above 10°C represents one degree day so, for example:

- 1 May: mean temperature 14°C; therefore 1 May = 4 degree days;
- 31 July: mean temperature 24°C; therefore 31 July = 14 degree days.

Degree days are measured in either Fahrenheit or Celsius, but it is important when comparing different regions to compare like with like, i.e. °C with °C, or °F with °F. The total number of degree days is then calculated. For example, in the USA, the coolest climatic region (Region 1) totals less than 2500 degree days (F), while the hottest (Region 5) has more than 4000 degree days (F). Based on this classification, suitable varieties are selected to match a region's climate. Some viticulturalists challenge the concept of degree days, regarding it as a simplistic, or inappropriate, measure of mesoclimates.

2.6 Impact of climate

Climate has a major impact upon the style of wines produced in any region. We might contrast a Riesling wine from a cool climate such as Germany's Mosel, with an alcohol content somewhere between 7.5% and 10% abv and a Riesling from Clare Valley, South Australia, with at least 12.5% abv. Generally speaking, as sugar levels increase during the ripening process, acidity levels fall. Consequently, the German Riesling is likely to have a higher acidity than the Australian wine, although it should be noted that the Australian producer is permitted to add acid during the winemaking process if considered desirable. Unless a derogation is given by the EU, as happened in the exceptionally hot year of 2003, acidification is not permitted in the cooler regions of Europe. Black grapes need more sunshine and heat than white grapes, to ensure the physiological ripeness of the tannins in their skins. It follows that in cool regions, grape production is predominately of white varieties, as exemplified by Alsace and the Loire Valley in France and the vineyards of Germany and England. A comparison of the average annual sunshine hours and temperatures of some selected production regions, in which very different styles of wine are produced, is shown in Table 2.1.

Vines perform at their best where there is a dormant period of about 5 months and a growing and fruit-ripening period of about 7 months. Essentially, sufficient warmth and moisture are required to be able to grow, produce and ripen grapes. It is, however, important that these conditions come at the right time in the vine's annual growth cycle.

Table 2.1 Comparison of annual sunshine and temperatures in selected wine regions

	Average annual duration of sunshine (h)	Average annual temperature	
		°C	°F
Mosel (Germany)	1400	10	50
Champagne (France)	1650	10.2	50.4
Bordeaux (France)	2000	12.7	54.9
Châteauneuf-du-Pape (France)	2800	13.5	56.3
Sonoma (California)	3000	14.8	58.6
Margaret River (Australia)	3000	16.4	61.5
Tulbagh (South Africa)	3300	17.3	63.1

2.7 Weather

Whereas climate is determined by geographical location and measured in long-term averages, weather is the result of day-to-day variation of those averages. The weather is the great variable in the vine's growing season and its production. Nearly every other major influence is more or less constant and known in advance. In the end, however, it is the annual weather conditions that can make or break a vintage. Weather not only influences the quality and style of wine produced in any given year but also can account for considerable variations in the quantity of production.

2.8 Climate change

Of all the topics in the wine world that provoke heated argument, the most pertinent is perhaps that of the impact, and the potential impact, of climate change. The quantity, style and quality of grapes produced result from a complex interaction among several natural factors, including climate. Different grape varieties have different climatic requirements, especially the average temperature required to fully ripen. However, if these averages are exceeded, there is a negative impact upon the quality of fruit, particularly the balance and complexity of the berries. In other words, the finest grapes for winemaking are grown close to the viticultural climatic edge for the individual varieties.

Climate has been changing since the world began, but many would argue that the changes in the last 30 years or so have been far more rapid than that endured by previous generations of fauna and flora. A 'medieval warm period' occurred in Europe from 950 to 1200 AD – this was perhaps the first 'golden age' for viticulture in England, although many English producers believe the second golden age is just beginning. However, in the world's wine regions, the rate of climate change in recent years has varied considerably, sometimes within relatively small or neighbouring production regions. In the last 35 years, the average temperature during the growing season of 1 April to 31 October in Bordeaux has increased by 2.64°C (4.75°F). In Napa, California, during the same period, the average has been only 0.94°C (1.69°F), and in neighbouring Sonoma, the average has not changed at all. It should be noted that there have been other periods when temperatures have been similar to those of the last decade, e.g. in the late 1940s and early 1950s in Bordeaux. Of course, one of the effects of higher temperatures during the growing season is to bring forward important events such as the flowering of the vine and *veraison*, the changing of the colour of black grapes from green to red. Harvesting time is also brought forward, perhaps having the advantage of bringing the fruit in before the onset of autumn rains. During the three decades from 1971 to 2000, the average dates of peak flowering, mid-veraison and commencement of harvest in the Bordeaux region were all getting progressively earlier. However, since 2001, with the exception of the early years of 1997, 2003 and 2011, the average dates remain unchanged. Table 2.2 shows average key dates in the growing season for the Merlot variety at five selected Bordeaux châteaux. Bearing in mind that in the years from 1985 to 2010, growers were progressively leaving fruit on the vine longer, in order to attempt to achieve phenolic ripeness, the harvest date figures in Table 2.2 perhaps understate the influence of climate change on the key dates.

Of course, climate change manifests itself in other ways than sunshine and heat. Rainfall patterns are changed, disastrous frosts can come during the spring when warmer temperatures have stimulated growth, and minimum temperatures in winter or spring may be higher or lower. For example, the average minimum temperature in South Australia's Barossa Valley has decreased by 2.9°C (5.2°F) in the last 30 years. Higher winter temperatures and shorter winters may reduce the productive life of vines, as a dormancy period promotes longevity.

Climate change may result in some of the greatest wine regions in the world losing their superiority within the next 30 years or so. It is possible that

Table 2.2 1970–2014 average key dates in the growing season for Merlot variety at five selected Bordeaux châteaux

Period	Peak of flowering	Peak of veraison	Start of harvest
1971–1980	17 June	22 August	1 October
1981–1991	13 June	17 August	28 September
1991–2000	3 June	9 August	21 September
2001–2010	2 June	9 August	21 September
2011–2014	2 June	8 August	20 September

the climate in the region of Bordeaux will no longer be suitable for the production of top-quality Cabernet Sauvignon grapes as soon as 2050. Countries and regions in which, to date, some varieties have struggled to ripen, e.g. England, may become home to some of the world's greatest wines. However, nature has a habit of reversing conditions, and projections by 'experts' can prove false. Speaking of the 2011–2014 growing seasons in Bordeaux, Professor Denis Dubourdieu, researcher, property owner and oenology consultant, says:

> The conditions changed our perception of climate. Since the 2001 vintage we had become accustomed to thinking we were in a period of global warming, that harvesting would be later and later, with increasingly ripe grapes, and we would no longer need to treat the vines. But this wasn't the case at all, and we were back to the old, familiar Bordeaux climate, with occasional rainy downpours.

CHAPTER 3

Soil

Soil is an important factor in the production of grapes for winemaking, and a variety of components influence the nature and quality of the soils. This chapter looks at the key factors for consideration in terms of soil suitability for viticulture, and the impact upon the styles of wine produced upon different soil types.

There is still some dissension over the impact that soil has on the quality of wines. The traditional Old World view, particularly held in some areas, is that vines only yield high-quality grapes if the soil is poor, and the vine is made to work hard for its nutrients. Many New World producers dispute this, believing that providing vigour is controlled, richer soils can yield good-quality grapes. The influence of soil upon wine quality is discussed very briefly here, and this topic will be developed further in Chapter 23. However, it is beyond dispute that a slate soil such as may be found in the Mosel in Germany, and illustrated in Figure 3.1, will produce a very different style of wine to the gravel of Bordeaux's left bank, as illustrated in Figure 3.2. or the schist of the Portugal's Douro Valley as illustrated in Figure 3.3.

3.1 Soil requirements of the grape vine

Many types of soil are suitable for successful vine growing, given the correct preparation. Essentially, the function of the soil is to provide anchorage, water, nutrients and drainage. Vines can and do thrive in the most inhospitable soils and unlikely sites. In considering the soil structure, both topsoil and subsoil have important roles: topsoil supports most of the root system including most of the feeding network. In some regions, such as Champagne, the topsoil is very thin,

Wine Production and Quality, Second Edition. Keith Grainger and Hazel Tattersall.
© 2016 John Wiley & Sons, Ltd. Published 2016 by John Wiley & Sons, Ltd.

Figure 3.1 Slate soil – Mosel.

Figure 3.2 Gravel soil – Bordeaux, left bank.

Figure 3.3 Schist soil – Douro.

and regular nutrient additions are required. The subsoil influences drainage and ideally enables the roots to penetrate deeply in search of nutrients and water.

Before planting a new vineyard, a full soil analysis should take place to examine the chemical and biological composition and balance. The pH of soils suitable for viticulture lies between 5.0 and an absolute maximum of 8.5, with an ideal range of 5.5–7.0. The pH of the soil has an impact upon the availability of nutrients and thus on the style and quality of wine produced. All other factors being equal, vines grown on a high-acid (low-pH) soil will produce grapes with a lower acidity than those grown on a low-acid (high-pH) soil. It is soil ion availability that aids or impedes acid retention in the grapes. Regular additions and adjustments to the soil, e.g. the addition of lime to acidic soils, such as those commonly found in South Africa, may be needed when the vineyard is established. Essential nutrients must be available to the vine, and saline soils are best avoided, since sodium chloride (NaCl), even in small quantities, severely reduces the availability of potassium (K).

The most important physical characteristics of different soil types are those that govern water supply, water retention and drainage – put simply, quality wine is not produced from poorly drained vineyards. However, much of the water retained in the soil is not available to the vine. The texture of the soil will depend on the proportions of gravel, sand, silt and clay, and affects drainage and the vine's ability to take up water, nutrients and minerals. Texture is very hard to adjust. Soil structure results from the type of aggregates (lumps of soil particles) and the spaces between them, and can be altered by preparation and vineyard management techniques, such as the addition of gypsum (calcium sulfate, $CaSO_4 \cdot 2H_2O$).

The key factors that need to be considered when planning and preparing a vineyard site are detailed below.

3.1.1 Good drainage

Like any other plant, the vine needs water and will probe deep to try and find it, so the best soil is one with good natural drainage and free from obstacles likely to impede the roots. There is a view that if the soil is poorly drained, the grapes may be inferior because the plant is taking up surface water from recent rainfall or irrigation, rather than the mineral-rich deep water. When preparing a vineyard, a drainage system should be installed if the natural drainage is inadequate.

3.1.2 Fertility

Although it is generally recognised that the vine thrives on poor soil, vines do require an adequate level of nourishment, and the addition of humus and organic matter may need to be done annually. In highly fertile soils, careful vineyard-management techniques are needed to restrict vine vigour and maintain balance between the vegetative and reproductive vine growth. The aim, of course, is to channel the vine's desire for reproduction into producing top-quality grapes.

3.1.3 Nutrients and minerals

Nitrogen (N) is an important nutrient requirement, essential for the vine's production of green matter. Again, balance is important so that excessive growth is not stimulated. Essential mineral requirements include potassium (K), calcium (Ca) and phosphorus (P). These should be present within accepted parameters, for both deficiencies and excesses can result in problems with vine health and grape and wine quality. In addition to the above minerals, trace elements such as magnesium, (Mg), sulfur (S), iron (Fe), manganese (Mn), zinc (Zn) and boron (B) are beneficial. Vine roots will seek out the areas richest in minerals, a process known as positive chemotropism.

3.2 Influence of soils upon wine style and quality

The style and quality of wines produced result from a complex interaction of all the natural factors, including climate and soil, and the decisions and work undertaken in the vineyard and winery. Even soil colour will have an impact upon grape ripeness: darker soils are warmer than pale ones. There has been considerable research into the impact that various soils have upon the wine in the bottle, and we examine this further in Chapters 23 and 25.

3.3 Soil types suitable for viticulture

We will briefly consider some of the types of soil found in various wine regions.

3.3.1 Limestone
Limestone is a sedimentary rock consisting mainly of calcium carbonate ($CaCO_3$). In limestone soils, the vine has to burrow its roots deep into the fissures to seek water and nutrients. This type of soil is valued in cool viticultural regions, for example in producing the great wines of Burgundy where the Jurassic limestone contributes to the wines' finesse.

3.3.2 Chalk
Chalk is a type of limestone that is very porous but has the dual qualities of good drainage while providing adequate water retention. The most famous chalk vineyards are perhaps those of the Champagne region of northern France, where two subtly different types of chalk, *Belemnita quadrata* and *Micraster*, provide the basis of most of the plantings. These soils are from the late Cretaceous geological period, being formed about 70 million years ago, and so are much younger than the limestones of Burgundy. In the Jerez region in southern Spain, the albariza soil is valued for its high chalk content. In the hot, semi-arid climate, the soil's capacity for absorbing and storing water over the winter period is invaluable, for very little rain falls during the summer months.

3.3.3 Clay
Clay is noted for poor drainage, but consequently is good for water retention. In some areas, clay is valued for its role as vineyard subsoil, for example in the Bordeaux districts of Pomerol and Saint-Émilion. However, clay soils are slow to warm up after the winter, thus delaying the start of the growing season, and are easily compacted. Soil compaction in vineyards, often caused by the passing of tractors and other machinery, must be avoided, as the soil will be deprived of oxygen, which is necessary for root respiration and potassium uptake. Thus, the ability of the roots to grow will be restricted.

3.3.4 Marl
Marl is a mixture of clay and limestone. In Burgundy's Côte d'Or, home to some of the world's finest Pinot Noirs and Chardonnays, there are many complex soil strata, including limestone, clay and marl. The Nebbiolo variety also loves marl soils, as demonstrated by fine Barolos produced in the hills around Alba in the Italian region of Piemonte.

3.3.5 Granite
Granite is a hard crystalline rock, rich in minerals, that warms up quickly and retains heat. Granitic soils provide good drainage and low fertility. The individual styles of the Beaujolais Crus are influenced by the variations in the granitic base soils. The Gamay variety, from which these wines are solely made, is naturally high in acid, and the acidic granite soils help counter this and produce balanced wines. Syrah is another variety that can excel on a granite base, e.g. the great wines of Côte Rôtie.

3.3.6 Gravel
The rocks prevent water from collecting, thus ensuring that vines have good drainage. The left bank of Bordeaux (Médoc and Graves) is famed for deep gravel banks, allowing the penetration of vine roots to several metres. During the day, the gravel will absorb the sun's heat, discharging it back onto the vines at night.

3.3.7 Greywacke
Greywacke is perhaps best known in much of New Zealand, including Marlborough, Canterbury and Central Otago. It is also found in some of the South African regions including Franschhoek, Sonoma County in California, a few parts of the Mosel and Spain's La Mancha. This grey sedimentary rock drains well and contains quartz. On a duller note, greywacke is used extensively as an aggregate for concrete making, roads and railway track beds!

3.3.8 Sand
Sand comprises grains of quartz, broken down from siliceous rock. This loose soil type, in which it is particularly difficult to store water and nutrients, is not easily cultivated. A positive quality of the soil is that the *Phylloxera* louse does not infest sand. Colares, on the coast to the north of Lisbon, is well known for its sandy soils.

3.3.9 Schist
Schist is a coarse-grained crystalline rock that is easily split into thin flakes. It is heat-retaining, ideal for the production of rich, robust wines, such as the Port wines of Portugal's Douro Valley. It contains high levels of the valuable minerals, potassium and magnesium.

3.3.10 Slate
Slate is a hard, dark and slab-like rock made of clay, shale and other elements. It has the advantage of holding moisture, warming up quickly and retaining its heat, then releasing it at night onto the vines. The blue Devon slate found in most of the Mosel region of Germany contributes to the racy character of the wines.

3.3.11 Basalt and other volcanic soils

These soils are rich in potassium and generally very fertile. The vineyards of Tokaji in Hungary, the Kaiserstuhl in the German region of Baden, Sicily's Etna, Campania's Vesuvius (Lacryma Christi) and the island of Madeira are all situated on volcanic soils.

3.4 Soil compatibility

Having considered the qualities of just some of the many soils in which the vine can be planted, it is important that the vine variety, clone and especially the rootstock planted be compatible with the soil (and climate). For example, chalk soils, although providing excellent drainage, can be low in fertility, resulting in low-vigour vines. When planting vineyards where there is a high level of active lime in the soil (e.g. most of Burgundy and Champagne), it is important to select lime-tolerant rootstocks; otherwise the vines may suffer from chlorosis. This is an iron deficiency, resulting in yellow leaves that have produced insufficient chlorophyll and therefore unable to manufacture sufficient grape sugars. An illustration of chlorosis-affected vines in Saint-Émilion is provided in Figure 3.4.

Figure 3.4 Chlorosis-affected vines.

3.5 Terroir

The influence of an individual growing location on style and quality is encapsulated by the French term 'terroir'. Terroir is essentially a combination of soil with all its various qualities, local topography (altitude, slope, aspect) and the meso- and micro-climates of the vineyard. As a result of a change in any of these factors, the vine can experience different growing conditions within just a few metres in the same vineyard. The concept of terroir is discussed in greater depth in Chapter 23.

CHAPTER 4

The vineyard

In this chapter, we examine site selection for a commercial vineyard; planting density; vine training systems; pruning methods; and how the canopy of shoots and leaves may be managed. The possible methods and timing of irrigation are also examined. The annual cycle of vineyard work is considered, and we detail the stages of development of the grape berry.

Views on viticulture will sometimes conflict, for what is the best theory and practice is not always universally accepted, and this dissention helps contribute to the contrasting styles of wine produced in different regions and countries. For example, one of the Old World concepts challenged by many New World producers is that in order to achieve good quality, it is necessary to restrict severely the quantity of grapes produced per hectare of land.

4.1 Vineyard location and site selection

The reasons for the historical situation of the world's great vineyards are complex. Producers in famed regions will wax lyrically about the outstanding terroirs of the greatest sites – aspect, soil, mesoclimate, etc. – as for example in the Grand Cru vineyards of Burgundy's Côte d'Or. Many great vineyards are close to rivers, and as we have already seen, these have an influence upon the vineyard climate. For example, Hermitage benefits from the cooling influence of the Rhône (and the Mistral wind), while the great vineyards of the Mosel and Rhein in Germany are warmed by heat reflection. However, it should be noted that the other great advantage of rivers historically was that they were the means of transport to the markets. With a strong market outside the region or country of origin, good wines would sell well at higher

Wine Production and Quality, Second Edition. Keith Grainger and Hazel Tattersall.
© 2016 John Wiley & Sons, Ltd. Published 2016 by John Wiley & Sons, Ltd.

prices. Higher prices mean more investment in the land, and export markets promote higher quality and the interchange of ideas between supplier and customer, so necessary to improve standards.

Today, when deciding on the location for a new vineyard, communication to the markets may be less of a consideration, but reasonable road access is still required. Indeed, in some countries, such as Australia, harvested grapes may be transported hundreds of miles to the winery. When investigating potential sites for new vineyards, the factors to be considered are many and complex. Clearly, the climate of region/site is a primary concern. As we have seen, grapes need a minimum of 1400 h of sunshine to ripen fully. However, the researcher will need to consider topography, aspect, orientation, mesoclimate, soil composition, suitability for the proposed grape varieties and rainfall. The availability of water may be crucial; not only for irrigation where desired and permitted, but also for use in sprinkler systems for protection where, in cool climates, there is a danger of frost. Availability of labour, the constraints of wine laws and any 'appellation system' as exists in Europe, together with site security, must also be taken into account.

4.2 Density of planting of vines

The number of vines planted per hectare will depend on a number of factors, including historical, legal and practical. In France, in much of Burgundy and parts of Bordeaux, including much of the Médoc, vines are planted at a density of 10,000 vines per hectare. The rows are 1 m apart, and the vines are spaced at 1 m within the rows. Mechanical working takes place by using straddle tractors, which are designed to ride above the vines with their wheels on either side of a row. A grower may state that by having such dense planting, vigour is kept in check, and the vines are slightly stressed, having to fight for their water and sending their roots down deep to pick up all the minerals and trace elements. Much of the available water goes into the structure of the plant, restricting the amount that will go into the grapes. This may result in more concentrated and better-flavoured juice. One of the problems that can arise with dense planting is that of soil compaction – the wheels of the straddle tractor follow the same path with each pass. In some other areas of Bordeaux, such as the Entre-Deux-Mers, the district between the Rivers Garonne and Dordogne, the plantings are less dense, some 4000 or so vines per hectare. Here, the soils are heavier, and historically two animals, rather than one, were needed to pull a plough, hence the wider spacing. So, historical influences live on simply out of practicality, but also there are many growers who are loathe to change traditional practices. However, in the vineyards of Bordeaux's left bank, prior to the arrival of *Phylloxera*, vines were planted at a density of 20,000 per hectare – even up to 40,000 in some cases. These vines were not in rows, as there was no mechanisation, and weeding was carried out by hand. Some growers are now returning to very dense plantings. These include the estates of 20 Mille (20,000 vines per hectare) and Dom. Léandre-Chevalier, where up to 33,333 vines per hectare are planted.

In other parts of France, plantings may be less dense, e.g. 3000 vines per hectare in Châteauneuf-du-Pape. In many New World countries, plantings might be as low as 2000 vines per hectare. Such density allows for the passage of large agricultural machinery, which growers might have been using prior to converting land to vineyards. Wide rows and inter-vine spacing allow sunlight into the vines. Open spacing also helps keep plenty of air circulating and reduces humidity – this can help prevent mildews and other fungal diseases of the vine. However, there are vineyards in the New World where the planting is just as dense as in France or Germany, for example parts of Aconcagua and Colchagua in Chile.

4.3 Training systems

Training and pruning are terms that sometimes cause confusion. Essentially, training combines the design and structure of the support system, if any, with the shape of the plant. Pruning is the management of this shape by cutting as required and is a means of controlling yield and maintaining fruiting.

The training system used may be based upon a number of factors, including historical, legal, climatic and practical. The ancient Romans trained vines up trees, with other crops growing underneath. The Greeks preferred low training systems in dedicated vineyards. Nowadays, there are several ways of classifying training systems, such as: by the height of the vines (low/high); by the type of pruning (cane/spur); and by the type of any support system used (trellis/pergola/free-standing bush).

4.3.1 Main types of vine training
4.3.1.1 Bush training
Bush-trained vines are either free-standing or supported by one or more stakes. Historically, this was the most widely practised training system in France, but bush-trained vines are not suitable for modern mechanised vineyard work, including trimming and mechanical harvesting. With a compact, sturdy plant, the system is particularly appropriate for hot, dry climates. The vine arms are spur pruned, i.e. just one or two buds are left upon spurs of the previous year's wood. If the bush is kept close to the ground, the vine will benefit from reflected heat, giving extra ripeness. If the leaves are allowed to shade the grapes, the system can produce grapes with pronounced green or

Figure 4.1 Bush vine, recently pruned, in Valencia.

grassy flavours, which may or may not be desired. Vineyard areas where bush training is practised include Beaujolais and parts of the Rhône valley. A recently spur pruned old Bobal bush vine in the Valencia region of Spain is illustrated in Figure 4.1.

4.3.1.2 Replacement cane training

Vines are trained using a wire trellis system for support. One or more canes run along the wires according to the particular system used. A popular method of cane training is the guyot system, in which one cane (single or simple guyot) or two canes (double guyot) are tied to the bottom wire of the trellis. Among the classic vineyard areas where the guyot system is practised are most of Bordeaux, including the Médoc, and the Côte d'Or in Burgundy. A diagram of a double guyot vine is shown in Figure 4.2. A photograph of unpruned vines in early spring is shown in Figure 4.3, a recently pruned guyot trained vine in Figure 4.4 and a guyot pruned vine following tying to the bottom wire in Figure 4.5.

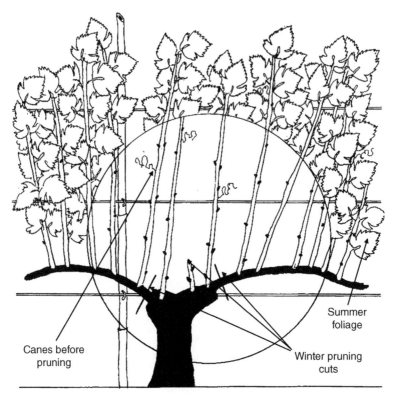

Canes before
pruning

Summer
foliage

Winter pruning
cuts

Figure 4.2 Diagram guyot vine showing pruning cuts. Source: Courtesy of Christopher Willsmore.

Figure 4.3 Unpruned guyot trained vines.

Figure 4.4 Recently pruned guyot trained vine.

Figure 4.5 Guyot trained vines following pruning and tying.

4.3.1.3 Cordon training

There are many variations of this system, but all have the principle that the trunk of the vine is extended horizontally along wires. The trunk extension (cordon) may be close to or high from the ground. Cordon systems are well suited to mechanisation, including mechanical pruning and harvesting. They are also appropriate in vineyards that suffer from strong winds, which would break the canes of vines trained in a replacement cane system such as guyot.

4.3.1.4 Vertical shoot positioning (VSP)

This training system requires four pairs of wires on a trellis. As the name implies, the vine's shoots are trained vertically. There are many variations of the system, including both replacement cane, and cordon and spur pruning. Cordon and spur is extensively practised in many New World countries, including Australia and Chile. Two cordons are trained in opposite directions along the bottom wire (which is approximately 1 m off the ground), and each of these will be allowed perhaps 10 or 12 spurs, which are pruned to one or two buds. The 'fruiting zone' will be 0.9 and 1.2 m above the ground. With the replacement cane system, practised in New Zealand, there are four canes, two in each direction, with one cane tied to the bottom wire and another to the next wire, which is approximately 15 or 20 cm higher.

Advantages of VSP training include reduced risk of fungal diseases, as the vines are not close to the ground, and the canopy is not dense, resulting in less humidity within the canopy. The fruit is contained in a compact zone, with much of the leaf area and the tips of the shoots higher. Thus, mechanisation, including mechanical harvesting, is facilitated. Figure 4.6 shows VSP vines after pruning in winter, and Figure 4.7 shows VSP trained vines in summer.

4.3.2 Other training systems

There are many other training systems. Some have been used since ancient times such as pergolas, and this system is still used in many areas. The pergola may comprise a complete overhead canopy, or a partial overhead system. An illustration of a pergola as used in Veneto in north-eastern Italy is shown in Figure 4.8. Others have been relatively recently developed or refined, including double hedge and lyre systems, as shown in Figure 4.9, and vertically divided canopies such as Geneva Double Curtain (GDC), Scott Henry Trellis (SH) and Te Kauwhata Two Tier (TK2T). Constant research on training systems is taking place in schools of viticulture.

4.4 Pruning methods and canopy management

Whatever system of training is used, the vine will need pruning to maintain the shape, manage the vigour and control the yield. The canopy of the vine consists of the parts above ground: trunk, stems, leaves, flowers and fruit. Good canopy

Figure 4.6 VSP vines after pruning.

Figure 4.7 VSP trained vines in summer.

Figure 4.8 Pergola in Veneto.

Figure 4.9 Lyre trained vines in Chile.

management is important because it will have an effect on the vigour and yield, and also help to reduce disease in the vine. Problems such as uneven ripening caused by shading of fruit and frost damage can also be contained.

4.4.1 Pruning methods

The main pruning takes place when the vines are dormant. Many growers wait until mid to late winter – in the northern hemisphere, January until early March. While pruning earlier can help stimulate the vine into earlier growth, growers fear die-back in the event of severe frosts. The pruning cuts away most of the previous season's growth. In the case of a replacement cane system, just one or two canes with perhaps four to eight buds each will be retained – in the EU, for Appellation Contrôlée (Appellation Protégée) wines, the maximum number of buds may be specified. Bush and cordon systems will be pruned to a number of short spurs with one or two buds each. The canes or spurs are wood that developed in the previous growing season. Grapes will only develop from the flowers that grow on the shoots from the previous year's wood.

As we have seen, yield management is a critical part of vineyard work. The yield of grapes to be harvested will depend on many factors including pruning, vintage conditions, local wine laws and winemaker's preference. If vines are pruned hard, the amount of grapes produced is reduced, and as we have seen, the conventional view is that this promotes quality. In France, the more illustrious and precise the appellation, the more restricted is the legally permitted yield. For example, the basic permitted yield for many Grand Cru (and several other) Burgundies is as low as 35 hectolitres per hectare (35 hl/ha). Many growers, especially in the New World, challenge this traditional view and claim that high yields can produce good quality. There are parts of Australia, heavily irrigated, where the juice that goes into some popular, inexpensive brands is produced at a rate of 200 hl/ha or more!

4.4.2 Canopy management

Good canopy management, control of the vines' foliage, is crucial to let plenty of sunlight into the vines, which promotes grape ripeness and quality. Canopy management begins with the winter pruning but continues through the growing season and may include shoot thinning, removing lateral shoots, shoot positioning, summer pruning and leaf removal. Open canopies are highly desirable to help prevent the onset of fungal diseases, which we will discuss in Chapter 5.

4.5 Irrigation

In New World countries, irrigation is extensively practised. In many regions, viticulture would not be possible otherwise, but it can also be used as a means of yield enhancement. The use of *flood irrigation*, where water is run over the

Figure 4.10 Furrow irrigation in Argentina. Source: Courtesy of Zuccardi Argentina.

entire surface of the vineyard, is regarded by many as inefficient, but proponents say that it encourages a large root system and helps to control certain soil pests. *Furrow irrigation* directs the water more specifically along the rows of vines with the benefit of less water usage. *Micro-jet irrigation* involves the vines being watered by tiny rotating nozzles resembling miniature garden sprinklers, placed above the trellis system. However, the system preferred by most is *drip irrigation*: a plastic pipe, often attached to the bottom wire of the trellis, dispenses a tiny amount of water directly to the base of each individual vine. Figure 4.10 shows furrow irrigation taking place in Argentina, and Figure 4.11 shows drip irrigation in New Zealand.

Deficit irrigation can be used as a means of vigour control – there is a view that vines, like humans, work best when slightly stressed! Irrigation should not take place in the weeks immediately preceding the harvest. However, in areas where irrigation is permitted, following the harvest, such an irrigation is commonplace to stimulate growth and replacement of plant energy.

Until recently, irrigation was generally not permitted in the EU, except in the case of young or experimental vineyards. However, in many regions, such as Alentejo in southern Portugal, irrigation is now commonplace. Although irrigation of vineyards in France was first officially allowed by a decree of 6 December 2006, in most of the classic regions it is still not permitted for Appellation Contrôlée (Appellation Protégée) wines unless the local laws specifically allow. Any irrigation must cease by 15 August. It should be noted that

Figure 4.11 Drip irrigation.

it is not usually required, but in drought years such as 2003, lack of an irrigation system and water results in severely reduced production.

4.6 The vineyard cycle and work in the vineyard

Wherever the location of the vineyard in the world, there are similar essential routine tasks that have to be carried out throughout the year. By way of example, let us consider the growing season in a cool climate, northern hemisphere vineyard.

4.6.1 Winter

Winter pruning is one of the most labour-intensive activities in the vineyard, and the skill of the pruner will impact upon the progress of the vines throughout the growing seasons. Following the pruning in January or February, the vines, if on a trellis system, need tying. Even if mechanical pruning has been used, there will be some further handwork to be done in cleaning up. If the guyot system is used, the canes will be secured to the bottom wire of the trellis system. Many growers leave this operation until March, when the sap is beginning to rise, and the canes are more flexible. Ploughing of the earth between the rows, or alternate rows, may take place at this time.

4.6.2 Spring
Planting of new vines generally takes place in the early spring, when the soil is beginning to warm up. Vines grow at temperatures above 10°C (50°F). The vine will normally bud in April, and any frost now could be disastrous, wiping out much of the year's crop. As we have seen when discussing climate, oil burners, wind machines or aspersion techniques may be used to try and prevent frost damage. The last is now the preferred technique in many areas.

In late April or early May, perhaps 2 weeks after budburst, any necessary spraying against mildews begins. Some growers spray routinely; others spray when their early warning system (e.g. a weather station) indicates that mildews might be expected. During May, the shoots will grow rapidly and will need placing between the trellis wires, a task that will continue in the following months. As the weather warms, lively new growth may need some canopy trimming.

Some growers like a weed-free vineyard; others encourage growth of certain weeds, or plant cover crops such as barley or legumes between the rows to help maintain soil balance. These should not be allowed to grow too high, perhaps 60 cm at most, as they would deplete the soil and create high humidity and thus encourage mildews. For the same reason, inter-vine weeds should be removed mechanically or by spraying. The area between the rows may be mowed or tilled according to the cultivation plan.

4.6.3 Summer
In late May or June, calm warm weather will encourage successful flowering. The vine's flowers, as illustrated in Figure 4.12, are hardly the most exuberant of flowers – the vine saves its beauty until the wine is in the glass. If the weather at flowering time is cold, windy or rainy, the vines may suffer from climatic coulure, the failure of flowers to develop into grape berries owing to their not opening or an otherwise unsuccessful pollination. Such weather conditions can also lead to millerandage, also known as shot berries or hen(s) and chicken(s), where some berries in a cluster remain tiny, while others develop normally.

By late June or early July, the fruit should have set, and the berries begin to swell. A little light rain from late June to early August is beneficial, for over-stressed vines will produce under-ripe fruit. During the summer, the tops of the vines may be trimmed, and leaves and entire shoots may be removed to maximise the vine's efficiency. Lateral shoots should always

Figure 4.12 Vine flowers.

be removed, as these are feeding on the energy of the vine. However, too much leaf removal will result in under-ripe grapes, for the leaves are the food factory of the vine. A formula to calculate the amount of leaf area required to ripen a given quantity of grapes was developed by Dr Richard Smart, an expert in canopy management. The leaf area required is between 7 and 14 cm^2 (dependent upon light and climatic conditions) per gram of ripe fresh fruit. During the summer, green harvesting or crop thinning may also take place, the aim being to channel the vine's energy into ripening a reduced crop. As the fruit begins to ripen, as detailed in 4.7 below, the grower may remove leaves from around the clusters to allow the grapes to bask in the sun. The afternoon sun is generally hotter than that of the morning, so it is sound practice to remove leaves only from the eastern side of the vine to minimise the risk of sunburn.

4.6.4 Autumn

As the grapes ripen, regular berry analysis will take place, and the time of harvest is decided. In a cool climate, the grapes may not be ripe until October. Many quality-conscious growers delay their harvest in the hope of a little extra ripeness, but if autumn rains set in, this decision may be regretted. Whether the grapes are picked by hand or machine will depend on a number of factors including, in parts of the EU, wine laws. Legislation states that grapes must be picked by hand for certain wines, including Beaujolais, Châteauneuf-du-Pape, Sauternes and Champagne. The method and timing of picking, along with the other viticultural decisions that a grower has to make, will impact upon the style and quality of the wine. There are advantages and disadvantages to each method, and each has its own characteristics. These will be discussed in Chapter 7. As the plants lose their leaves, vineyard maintenance will take place with general tidying, stakes and wires replaced, etc. In late autumn, fertiliser and nutrients may be applied to the vineyard and soil structure adjustments, e.g. the addition of gypsum (calcium sulfate) may take place.

4.7 Grape-berry development

During the period from fruit set to harvest, the growth and development of the grape berries may be considered in three phases:

- Phase 1. Following flowering, berry formation, including production of the seed, takes place. There follows a period of rapid growth, largely achieved by cell division. The grapes accumulate tartaric and malic acids, and tannins in the seeds and skins, generally at much higher levels in red varieties. Owing to the chlorophyll content, the berries are green and remain firm during this phase, which may take 55–60 days or so.
- Phase 2. This is often called the 'lag phase' and may last for 5–10 days. Grapes grow very little during this period. They still contain virtually no sugar. It is a convenient time for the grower to estimate the final crop, which, barring weather or other disasters, will be approximately double the weight of the berries at the lag phase.
- Phase 3. This starts with *veraison*. This is the beginning of ripening and, in the northern hemisphere, will probably take place in August. Sugar in the form of sucrose is transported into the berry and then hydrolysed into its constituents of fructose and glucose. Berry growth is now mainly due to the expansion, rather than the reproduction, of the cells. White berries become more yellow or translucent owing to the formation of *flavones*, and black grapes take on their colour owing to the formation of *anthocyanins*. The amount of malic acid in the berry falls considerably, especially in hotter climates, and tannins, particularly in the seeds, also decline. The development of aroma and flavour compounds, and the precursors of other compounds that will become volatile aromas as a result of the vinification process, begins fairly late in this phase. Such development begins about 30 days or so after veraison.

CHAPTER 5

Pests and diseases

As with any other plant, the vine is susceptible to many types of pest and disease. The most important are covered in this chapter. We consider how these may be prevented or controlled and the impact they may have upon the vine, the crop of fruit and the resulting wine quality.

Some pests, such as parasites, may derive their food from the host without causing too much harm. Others can be devastating. Diseases usually attack during the growing season and, unless controlled, can have a drastic effect on grape quality, quantity of fruit produced and health of the vine. Consequently, growers need to be vigilant and face a constant battle. Most vine diseases need heat, humidity or both, to flourish. Depending upon the season's weather, regular sprays for prevention or treatment may be required. However, rather than routine preventative spraying, the environmentally conscious producer utilises an integrated pest-management programme. This, together with organic and biodynamic regimes, will be discussed in Chapter 6.

5.1 Important vineyard pests

The following list of vineyard pests is far from exhaustive but includes many of the most serious. Some pose a problem in just a few regions within countries; others are present throughout the viticultural world.

Wine Production and Quality, Second Edition. Keith Grainger and Hazel Tattersall.
© 2016 John Wiley & Sons, Ltd. Published 2016 by John Wiley & Sons, Ltd.

5.1.1 Insects, mites and worms
5.1.1.1 *Phylloxera vastatrix* (also known as *Phylloxera vitifoliae* and now, owing to reclassification, more properly known as *Daktulosphaira vitifoliae* or *Viteus vitifoliae*)
This is the most serious of all pests and totally destructive, quickly killing vines, as we saw in Chapter 1. The only reliable solution is to use only grafted vines that comprise the required variety of *V. vinifera* grafted onto an American species rootstock or a hybrid thereof.

5.1.1.2 *Margarodes vitis*
This is the most important of several species of *Margarodes*. It is present across large parts of South America, being the major pest of the grape vine in Chile, and also found in South Africa. It was first recognised around 30 years ago. Like *Phylloxera* (with which it can be confused), *Margarodes* feeds on vine roots and results in stunted growth, leaf discolouration, leaf death and low or no crop. Over time, vine death is inevitable. However, if the vines are encouraged to root to a depth of 3–4 m, a depth at which the pest cannot survive, the vine may continue to live but with stunted growth. Heavy rainfall or flood irrigation can reduce pest numbers. Some producers are now grafting to try and combat the pest, but American rootstocks may not be as resistant as they are to *Phylloxera*. *Margarodes* can have the positive effect of restraining vine vigour.

5.1.1.3 Grape moths [e.g. pyrale (*Sparganothis pilleriana*), cochylis (*Eupoecilia ambiguella*), eudemis (*Lobesia botrana*) and light brown apple moth (*Epiphyas postvittana*)]
The caterpillars of these insects attack the buds of the vine and flower clusters in spring and early summer; later, the grapes are damaged, allowing disease to set in. Reproduction of the moths may be prevented by 'sexual confusion', as detailed in Chapter 6.

5.1.1.4 *Drosophila*
Spotted winged drosophila (*Drosophila suzukii*) and common fruit fly (*Drosophila melanogaster*) can be responsible for the onset of sour rot, imparting a distinct vinegar smell to the grapes.

5.1.1.5 Nematodes
Microscopic worms can feed on the root system, causing a variety of problems for the vine, including water stress and nutritional deficiency. There are over half a million species of nematode! The worms can spread virus diseases through the vineyard. If new land is to be planted, the soil should be carefully checked for the worms prior to planting – land that has previously borne potatoes or carrots is best avoided, as these crops host root-knot

nematodes. In areas where nematodes may be a problem, growers should choose nematode-resistant rootstocks.

5.1.1.6 Spider mites
These belong to the *Tetranychidae* family and are mites rather than true spiders, having no separate thorax and abdomen. Species include the red mite and common yellow mite. They infest the leaves and restrict growth.

5.1.1.7 Grape erineum mites also known as grapeleaf blister mites (*Colomerus vitis*)
These overwinter in buds and live and feed on the underside of leaves in summer, causing galls and raised blisters that affect the growth of young vines but otherwise do not have a major impact on quality or cause economic losses.

5.1.1.8 Rust mites (*Calepitrimerus vitis*)
These belong to the same species as grape erineum mite but are generally found on the upper surface of the leaves. They cause a reduction in yield. It is interesting to note that mites were largely unknown in vineyards before the 1950s, when the introduction of certain insecticides destroyed the previous biological balance. Treatments with sulfur may be effective.

5.1.1.9 European hornets (*Vespa crabro*)
These large yellow and brown wasps have a love of all fruit juices, and the pulp of very ripe grapes presents an easy source. In some countries, including Cyprus, Israel and Lebanon, they can pose a huge problem, and unless controlled, crop losses can be over 40%. Locating and destroying the nests is the only effective method of damage prevention.

5.1.1.10 Vine weevils
The adult weevils drill holes into the canes and lay eggs. A legless grub then tunnels along the canes, seriously reducing the strength of the vine.

5.1.2 Animals and birds
5.1.2.1 Animals
These include rabbits, hares, deer, wild boars, baboons and kangaroos, which dig, burrow or gnaw, and all can be very destructive. Rabbits and hares can strip bark and leaves, while larger animals are partial to the grapes.

5.1.2.2 Birds
Birds can be a serious problem during grape ripening, stripping vines of fruit, especially where vineyards are fairly isolated. Birds are selective and begin their destruction as soon as the grapes start to ripen. Some growers choose to remedy this by netting their vines or using audible bird scarers.

5.1.2.3 Snails

Snails are mainly a problem in early spring, especially when conditions are wet. They will eat buds and foliage but also encourage birds. However, this is a time when some birds can be beneficial – many birds feed on pests such as snails, and ducks and guinea fowl may be employed for this purpose.

5.2 Diseases

The vine and its fruit can succumb to many diseases, the most important of which are considered below.

5.2.1 Fungal diseases
5.2.1.1 *Botrytis cinerea*, also known as (*Botryotinia fuckeliana*)

This fungal infection may be welcomed or otherwise, depending on circumstances. *Botrytis cinerea* belongs to the Sclerotiniaceae family of the Ascomycetes class (which has over 1600 genera and 64,000 species, including yeasts, truffles and penicillium).

In the form of grey rot, *Botrytis cinerea* is most undesirable. It thrives in wet, humid conditions on vigorously growing vines. It can affect buds and young shoots, turning them brown. Later in the season, it can form grey mould patches on the leaves and infect flowers. It may then become dormant until the grapes are developed. White grapes will turn brown, and black grapes may take on red tones. The grapes are covered in a grey or grey/brown mould, and berries can split. Grey rot will result in off-flavours in wine, partially owing to the modified chemical composition of the grapes. However, its effect on black grapes is greater than white because it causes loss of colour, tannins and flavour. Outbreaks at harvest time (often induced by rain) can wreck a vintage – there will be a substantial loss in yield and a serious impact upon quality. An illustration of grapes attacked by *Botrytis* is shown in Figure 5.1.

In its benevolent form, *Botrytis cinerea* occurs as 'noble rot' on some varieties of white grapes when certain climatic conditions occur. The ideal conditions are damp, misty early autumn mornings giving way to very warm, sunny afternoons. Some of the great sweet white wines of the world are produced from affected grapes. More information on *Botrytis cinerea* in the form of 'noble rot' is given in Chapter 14.

5.2.1.2 Powdery mildew, also known as *Oidium tuckeri* (*Uncinula necator*)

This fungal disease reached the vineyards of Europe from North America between 1845 and 1854. It causes dull, pale grey spores to grow on leaves of the vine, and it can overwinter inside the buds. Eventually, the grapes will split and shrivel. In the case of severe infection, the entire harvest will be lost. The disease thrives in mild, cloudy conditions and especially where there is a

Figure 5.1 *Botrytis*-affected grapes.

dense, shady canopy of leaves. It may be treated by dusting the vines with sulfur powder, a contact method that works as both a preventative and an eradicant. Alternatively, systemic treatments are available.

5.2.1.3 Downy mildew, also known as *Peronospera* (*Plasmopara viticola*)

This is another fungal disease that came to Europe from North America and was first noticed on vines in the Bordeaux region in 1878. It had probably been carried on vines being imported as grafting stock to combat *Phylloxera*. The disease thrives in warm, wet weather, common to many North European wine regions. At least 10 mm of rain in a 24 h period is necessary for there to be an outbreak. Dense white growth forms on the underside of young leaves, which may shrivel and drop off. When attacking the grapes, it causes the berries to shrivel up and turn leathery. Affected grapes show either grey or brown rot. It may be treated by contact or systemic methods: a common contact treatment is spraying the vines with 'Bordeaux mixture' – copper sulfate, lime and water. Vineyards thus treated often show a blue hue, and the sprays even give a blue tint to pebbles or gravel on the soil surface. Some strains of downy mildew have developed resistance to some systemic treatments.

5.2.1.4 Black rot (*Guignardia bidwellii*)

This is yet another disease that was carried to Europe from the United States, being first noticed in France's Hérault Département in 1885. Infected grape berries shrivel and dry up, having turned blue-black in colour. Bordeaux mixture is an effective treatment.

5.2.1.5 *Black spot*, also known as anthracnose (*Elsinoe ampelina*)

This is a cryptogamic fungal disease that originated in Europe. It needs warm, humid conditions to thrive. It can attack young foliage and manifests itself on grape berries as small black dots. Eventually the berry splits. Copper treatments are also effective for anthracnose.

5.2.1.6 Eutypa dieback (*Eutypa lata*)

This fungal disease usually attacks older vines, entering through pruning wounds, and results in vine arms dying. The only real treatment is cutting back infected vines as soon as the disease is noticed.

5.2.1.7 Esca (also known as black measles on account of the spotting on the grapes of diseased vines)

This disease is probably the most serious threat in the wine world today. Infected vines initially suffer loss of foliage and stunted growth, but vine death can occur very quickly. Fungal species including *Phaeomoniella chlamydospora*, *Phaeoacremonium aleophilum*, *Phaeoacremonium inflatipes* and *Phaeoacremonium chlamydosporum* are known to be involved and generally infect the vines via pruning cuts. It is reported that in 2014, the disease affected 13% of French vines. Although the disease may be treated with sodium arsenite, the chemical is banned in the EU and elsewhere, as it is carcinogenic.

5.2.1.8 *Phomopsis* (*Phomopsis viticola*)

This is a particular problem in New Zealand but found in other cool regions. This disease spreads in wet springs and manifests as black spots surrounded by yellow rings on leaves. Black cracks appear at the base of shoots, which can then break off. Maintaining good air circulation around the vines by utilising open canopies is perhaps the best means of prevention.

5.2.2 Bacterial diseases

5.2.2.1 Pierce's disease

Prevalent in California, this bacterial disease caused by the bacterium *Xylella fastidiosa* is transmitted by small winged insects called 'sharp-shooters', which hop from leaf to leaf, feed on the vine and inject the organism. The bacterium affects food crops other than grapes, e.g. lucerne, and is commonly associated with reeds and other water vegetation. There is no treatment for infected vines, so minimising the insect numbers is crucial. The glassy-winged sharpshooter (*Homalodisca vitripennis* also known as *H. coagulate*) spreads infections particularly rapidly, feeding on the vines even in winter dormancy. Infected vines may have delayed or non-existent budburst, discoloured leaves and dry grape clusters. Vines may die within a few years of infection.

5.2.2.2 Bacterial blight (*Xylophilus ampelinus*)

This disease causes shoots to wilt and dry up, and can occasionally lead to crop loss. There is no treatment that destroys the bacteria, but disinfecting pruning tools and spraying of pruned vines with Bordeaux mixture can help prevent spread.

5.2.3 Virus diseases

The visually attractive red vine leaves seen in many vineyards in autumn can be a sign of virus-ridden vines. The University of California at Davis has identified over 55 virus and virus-like diseases of the grapevine. This type of disease is spread in several ways, including phloem-feeding insects such as leafhoppers (which are strongly attracted to yellow colours), but mainly by taking cuttings from infected plants. Once the need for grafting rootstocks to combat *Phylloxera* had been established at the end of the nineteenth century, it became critical to check that both the rootstock and scion used were virus-free. Grafting doubles the risk of virus infection as two different plants merge. The more common viruses include fan leaf (court-noué), yellow mosaic, flavescence dorée, fleck, leafroll and corky bark. An illustration of vines in South Africa infected with leaf roll virus is shown in Figure 5.2. Viruses do not usually kill the vine but gradually reduce growth and yield. As there is no

Figure 5.2 Virus-ridden vines in South Africa.

cure for viruses, prevention is the only answer. Nurseries supplying growers must ensure that material is virus-free. One way of ensuring this is to use heat treatment on original mother-garden material. In order to keep their vineyards free of viruses, growers must buy certified stock.

5.3 Prevention and treatments

From all points of view, prevention of pests and diseases is obviously better than cure. This may start at the site-selection stage by, for example, avoiding areas where there are known pests in the vicinity or where there is high humidity. It continues by only buying certified virus-free grafted vines from approved and regulated nurseries, and ensuring that all tools, implements and machinery are clean and pest-free, especially when moving them from vineyard to vineyard. During the annual cultivation cycle, care must be taken when pruning, to ensure that all cuts are clean. The avoidance of dense leaf canopies is key to minimising the danger of fungal diseases.

Spraying at optimum times in the growing season is carried out to prevent or treat diseases and control weeds. Tractors pulling spray machines are usually used for this, but if the terrain is very steep, then spraying has to be undertaken manually. Small planes or helicopters are an alternative, but there are always concerns about spray drift. At the time of writing, the use of drones in vineyards has been limited to gathering data – in Bordeaux, the first drones appeared in the district of Saint-Estèphe in the spring of 2013, monitoring the health of the vines. Tests are taking place in California to determine their suitability for spraying. Spraying is a costly business. Spray treatments may broadly be categorised as either contact, where the treatment remains on the surface of the leaves and other areas, or systemic, in which the agent enters the system of the plant and is freely transported through the tissues. The first spraying, perhaps using lime–sulfur (contact), may start in the spring as the buds swell and soften, and regular preventative sprays may continue throughout the season. However, as we shall see in the next chapter, more environmentally sensitive approaches to this and other matters of grape production have developed in recent years.

CHAPTER 6

Environmentally sensitive vineyard practices

In this chapter, we consider the adoption of environmentally sustainable vineyard practices. We look at integrated pest management (IPM), organic production and the fast-growing move by many growers into the holistic world of biodynamics.

The days when grape farmers were held hostage to the agro-chemical salesman are perhaps forever behind us. An ever-increasing number of growers proudly expound their organic credentials, and most of the remainder claim to be practising 'la lutte raisonnée', 'integrated pest management', 'sustainable viticulture' or whatever title they choose to use. During the 1960s, 1970s and 1980s, it seemed that you could not enter a vineyard without seeing routine preventative spraying, or applications of yield 'enhancing' chemicals. Now, cover crops abound, and the gentle hum of 'beneficial' insects is the main sound to be heard. And yet, for growers who embrace sustainability, the standards of vineyard hygiene are perhaps higher than ever.

6.1 Conventional viticulture

What is termed conventional viticulture is perhaps a misnomer, as the methods adopted only became commonplace with the development of vineyard machinery and the availability of NPK fertiliser, and particularly with the synthetic treatments introduced and aggressively marketed in the years after the Second World War.

Typically, conventional viticulture involves a monoculture. The inter-row earth is ploughed and maintained weed-free by the use of herbicides. There are regular additions to the soil of nitrogen, and chemical nutrients and

Wine Production and Quality, Second Edition. Keith Grainger and Hazel Tattersall.
© 2016 John Wiley & Sons, Ltd. Published 2016 by John Wiley & Sons, Ltd.

micronutrients are applied to the vines using foliar sprays. Routine preventative sprays of pesticides and anti-fungal agents are the norm – up to 20 times per year in some regions. However, when spraying is necessary, instead of cannon spraying it is possible to use a precise jet machine, or one with panels that collect and recycle surplus liquid, thereby reducing the volume used and chemical residues in the soil.

Opponents claim that conventional viticulture involves utilising sprays that kill anything that is harmful to the vine but, in doing so, often also kill the natural predators to vine pests. There can be little doubt that many treatments also kill microbes that surround the root zone that help the vine fix nutrients from the soil. Regular use of herbicides and insecticides creates a tolerance to them, and stronger concentrations or new treatments become necessary. The vine also develops a dependence upon such treatments, so simply withdrawing them will lead to an array of problems. Accordingly, conversion to more environmentally friendly methods requires consultation, advice and detailed planning.

6.2 IPM

IPM in the vineyard is also known as 'la viticulture raisonnée' and 'la lutte raisonnée' (the reasoned struggle).

In 1991, the Centre Technique Interprofessionnel de la Vigne et du Vin (ITV) published *Protection Raisonnée du Vignoble* (ITV, 1991). For many years this was both bible and practical guide to thousands of growers needing to fight pests and maladies of the vine, be they cryptogams, bacteria and mycoplasmata, or viruses. An IPM programme may not totally eliminate the problem but aims to control to the extent that there is no economic damage, or negative impact upon quality.

Conversion to IPM is reasonably straightforward. The grower considers the interaction between pests, diseases, vines, the environment and available methods of control. Possible problems are identified, the various prophylactic methods are considered, and the risk of the problems occurring is estimated (this is the real key to success). Treatments are then undertaken only as necessary and mindful of the wider consequences. Detailed records should be made.

In many regions, networks of weather stations have been established, producing daily reports warning subscribing growers if conditions are such that outbreaks of particular diseases are likely. Prevention might be achieved simply by de-leafing and opening up the canopy, but if spraying is necessary, it will be timely and therefore most effective. For the prevention of primary infections of cryptogamic diseases, the grower keeps a watchful eye out for 10:10:24 conditions, i.e. there is at least 10 mm of rainfall, when the

Figure 6.1 Sexual confusion ampoules.

temperature is above 10°C (50°F) during a 24 h period. Unless sunshine dries off the foliage, immediate spraying is necessary.

The problem of some insect pests, particularly grape moths, can be reduced by hanging small capsules containing pheromones spaced at 2 m intervals on the trellis wires, as shown in Figure 6.1. These create 'sexual confusion' in the male and thus prevent breeding. Beneficial insects are unaffected.

As there is no obligatory checking process, or accreditation of producers practising IPM, it is perhaps hard to distinguish the dedicated adherents from those whose environmental efforts are more in words than deeds.

6.3 Organic viticulture

Of course, all farming was once 'organic', but the name was given as recently as 1946, by J. I. Rodale. He created a demonstration farm in Pennsylvania and was the founder of Rodale Press. However, as little as 30 years ago, there were few producers of organic wine grapes of high quality. Today, adherent growers have shown that organically grown grapes can actually be of higher quality and sometimes even produced at lower cost than those grown conventionally.

Of course, organic farming eschews the use of pesticides, herbicides, fungicides and synthesised fertilisers. The use of cover crops is considered essential. These are secondary crops planted between the vine rows and have many purposes including, improving soil structure, increasing nutrient levels with the decaying of organic matter and also increasing micro-organisms and beneficial worms. Cereals such as oats or barley provide excellent green manure. Legumes such as faba beans, field peas, vetches and lupins help fix nitrogen from the atmosphere. Flowering plants are specifically included to attract the vine pests' natural predators. Cover crops planted in a vineyard in Trentino, Italy are shown in Figure 6.2.

Organic vineyard scenes can be particularly bucolic, with geese, ducks, strutting turkeys and even alpacas – all for a purpose. The fowl and alpacas eat surface pests and keep down cover crops. Alpacas are blessed with the attribute of soft hooves, thus reducing soil compaction. From the creation of a biodiversity project, it will take 4–5 years for the positive effects to become apparent. Many organic and biodynamic growers are now reducing their reliance upon

Figure 6.2 Cover crops in New Zealand.

tractors and reintroducing animals (horses and oxen) for working the fields, returning to a scene that was commonplace a century ago.

Organic viticulture in regions with warm, dry summers is less challenging than in cooler, more humid areas. Nevertheless, there are always pests that pose particular problems. At one time countered with chemicals, these can now be dealt with using natural ingenuity. By way of example, we will consider some particular problems and solutions as dealt with by some organic producers in Chile.

The burrito (meaning 'little donkey') is a nasty mite that, when given a chance, munches its way through the leaf structure, starving the plant of nutrients and stunting growth. The organic defence is a binding of plastic material with a sticky surface that is wrapped around the vine's trunk about half a metre from the ground. When the burrito tries to climb up to its leafy lunch, it gets stuck or turns around and retreats to the earth. Some producers employ a gaggle of geese that eat burritos and anything else they can bury their beaks into as they patrol the vineyard and fertilise the vines. The red (spider) mite is dealt with by the introduction of a larger white spider *Neoseiulus californicus* (harmless to vines) that feeds on the red mite. Grass plantings between the rows of vines complete the management by giving the white spiders something to munch on. Good vineyard hygiene is particularly necessary in organic production. The problem of escolito, an insect that lives on the canes, is reduced by ensuring that all pruned material is taken away and crushed. Wasps are countered by hanging used plastic soft-drink bottles into which holes have been punched. The sugary residue attracts the wasps, which are then trapped. Powdery mildew is countered with a spray of sulfur every 10 days before veraison (sulfur is a permitted contact treatment), and *Botrytis cinerea* (grey rot) is reduced by sprays of liquefied grapefruit, as shown in Figure 6.3.

This is perhaps a convenient point to note that prior to 1 August 2012, EU Regulations did not permit use of the term 'organic wine' per se, although 'wine made from organic grapes' was allowed. However, framework regulation EC No. 834/2007, followed by detailed regulation EU No. 203/2012, provides for use of the term 'organic wine', provided the producer satisfies all of the criteria.

6.4 Biodynamic viticulture

There is little in the world of viticulture that arouses such controversy, scepticism, hostility or passionate advocacy as the topic of biodynamic farming. Many scientists, including some experts in viticulture, consider biodynamics as 'unscientific tosh', indeed 'black magic'. However, producers of some of the world's greatest wines practise 'biodynamie'. In Bordeaux, these include Château Pontet-Canet (Pauillac), Château Latour (Pauillac) in part (some 24 ha of the estate's 47 ha l'Enclos vineyard) and Château Climens (Sauternes). In

Figure 6.3 Liquefied grapefruit anti-botrytis spray.

Burgundy, the roll-call of brilliant biodynamic producers is headed by Domaine de la Romanée Conti (Vosne-Romanée) and Domaine Leroy (based in Meursault) and Domaine Leflaive (Puligny-Montrachet). Elsewhere in France, flagship biodynamic producers include Domaine Bott-Geyl, Zind Humbrecht and Marcel Deiss (Alsace), Chapoutier (Rhône), and Domaine Huet and Coulée de Serrant (Loire). Sixty-five of the 240 ha owned by Champagne producer Louis Roederer are farmed biodynamically. Fine German biodynamic estates include Dr Bürklin-Wolf (Pfalz). In Margaret River in Western Australia, the Cullen Estate is totally biodynamic. In New Zealand's Central Otago, Felton Road and Quartz Reef are top-class biodynamic producers. The most famous biodynamic producer in Oregon is Beaux Frères, who have the world's most influential wine critic, Robert Parker Jr, as a partner. Few would dispute that wines from all the above, and many other biodynamic growers, can be of the highest quality, but the question remains: is the quality because of the biodynamic production, or is it that the producers totally understand their terroir and their vines, take so much care and give every attention to detail?

6.4.1 Rudolf Steiner

The birth of biodynamic farming stems from a series of eight lectures entitled 'Spiritual Foundations for the Renewal of Agriculture' given to a group of farmers in June 1924 by Dr Rudolf Steiner. He died a year later, but the term biodynamic, which translates to 'life forces' was not coined until after his death. Biodynamism is a holistic approach, which holds that plants are energised not just by the soil but by the air, terrestrial cycles and those of the sun, moon, planets and stars. Of course, synthetic fertilisers, herbicides and pesticides are not used. It should be noted that Steiner formed his ideas before synthetic, carbon-based pesticides were invented.

Biodynamic producers enhance the soil and the vines by applying special preparations, at the appropriate time in the natural cycle. The mixtures will be added on the designated day – there are 'leaf days', 'root days', 'flower days' and 'fruit days' according to the passage of the moon – and at the 'correct' time of the day.

6.4.2 Biodynamic preparations

There are nine biodynamic preparations, numbered 500–508. The two key preparations, 500 and 501, are detailed below, along with the reasons for their use, according to the biodynamic principles. The other preparations, 502–507 are listed, but for more information the reader is referred to the Bibliography.

6.4.2.1 Preparation 500 – horn manure

This is perhaps the key preparation for biodynamic farming. Dung is collected in autumn from fresh cow pats from a lactating cow. It is filled into cow's horns, which are buried (open end down) in a pit at a depth of 40–46 cm to overwinter. After 5 or 6 months, at the spring equinox, the horns are dug up and the contents combined with rainwater at a typical concentration of 60 g of manure to 30 or 40 litres of water. The mixture is dynamised for an hour by stirring first in one direction, until the vortex formed reaches the bottom of the container, and then in the other direction. This creates chaos and puts air and life back into the water. The quantity obtained from one horn (60–120 g) is sufficient for spraying on 1 ha of soil. The preparation is sprayed onto the earth two to four times a year and improves soil structure, increases microbial activity and gives the vines earth forces.

6.4.2.2 Preparation 501 – horn silica

This is ground silica put into a cow's horn and buried for the summer months. After digging up, a few grams is added to rainwater and dynamised for an hour as described above. The mixture is sprayed onto the vines in the early morning two to four times in the growing season to add light forces, improve photosynthesis, help disease resistance and grape ripening.

6.4.2.3 Preparation 502 – yarrow

This comprises yarrow flower heads, overwintered then filled into a red deer stag's bladder, which is then hung out the following summer. The bladder is buried 30 cm under the ground before the next winter and dug up in spring. This is then added in minute amounts to compost (as little as 5 ml/10–15 tonnes) along with other biodynamic preparations, to help trace elements in the soil become more available to the plants.

6.4.2.4 Preparation 503 – chamomile

Chamomile flowers are filled into a cow's intestine and overwintered in the ground. This preparation is used in tiny amounts as above to break down compost, vitalise the vines and prevent excessive growth.

6.4.2.5 Preparation 504 – stinging nettle

Cut nettles are buried underground for a full year to absorb the influence of all four seasons. This regulates iron in the soil and attracts magnesium and sulfur.

6.4.2.6 Preparation 505 – oak bark

Ground oak bark is filled into the skull of a goat, sheep, pig or horse and buried over winter in damp earth and then dried in glass jars. It is added to compost to add 'living' calcium to the soil and help the vine resist fungal diseases

6.4.2.7 Preparation 506 – dandelion

Dandelion flowers are filled into a cow's mesentery and overwintered under-ground. Added in minute amounts to compost, it brings cosmic, life forces.

6.4.2.8 Preparation 507 – valerian

Valerian flowers are collected on a summer morning. Ground and pressed with a little water into a tea, its use is to concentrate phosphorus in the soil.

6.4.2.9 Preparation 508 – horsetail decoction

The horsetail plant has the highest concentration of silica of all plants. Dried plants are added to water and simmered for up to an hour. It is then diluted and used as a spray to keep fungus spores in the soil rather than on plants. Figure 6.4 shows a series of pits containing biodynamic preparations.

Both organic and biodynamic producers are allowed to use copper sulfate and sulfur in the vineyard (and sulfur dioxide in the winery, albeit at a lower legal limit than that generally permitted), although some producers are having success with substituting vineyard sulfur with milk or whey, which counters powdery mildew. Both copper and sulfur are considered essential to life. However, copper is toxic, and continued use builds up levels in the soil, giving ammunition to those who question the environmental impact of some per-mitted organic and biodynamic practices.

Figure 6.4 Biodynamic preparation pits.

6.4.3 Certification

There are several certification bodies for organic producers, who thus have a choice as to which of the organisations they may subscribe. One organisation operating in and from the UK is the Soil Association, which is involved in the definition of a variety of organic agriculture certification schemes, as well as the auditing of producers to verify compliance with the schemes. As a form of organic agriculture, the interests of those involved with biodynamic agriculture are represented by Demeter-International e.V., which was founded in 1928 and is the main certification body for biodynamic farmers. Demeter has certification organisations in most parts of the world except, at the time of writing, sub-Saharan Africa and Australia. This organisation operates certification schemes for all kinds of biodynamic agriculture. A rival organisation, Biodyvin (Syndicat International des Vignerons en Culture Bio-Dynamique) was founded in 1995 and only operates in the field of viticulture. It now has over 100 members, nearly all in France.

6.5 Natural wine

Recent years have seen the birth and growth of a so-called 'natural wine' movement. The expression has no legal definition, and there is no additional regulation or certification of producers other than that for certified organic or biodynamic status. A somewhat loose definition, given by Isabelle Legeron MW, is that natural wine must be 'made from grapes organically or biodynamically farmed, harvested by hand, and have nothing added or taken away in the winemaking process.' High-tech techniques are eschewed. Proponents see wine being returned to its roots, but the consumer may sometimes need to adjust their palate, which has been conditioned tasting wines corrupted by additives and processing aids.

CHAPTER 7

The harvest

The harvest, when the grapes are picked and transferred to the winery, is a crucial time during the winemaking cycle. The grower/producer/winemaker must make the important decision as to the ripeness and sanitary condition of the grapes and when to pick. In this chapter, we consider the timing and various methods of harvesting, including the advantages and disadvantages of each method.

In the northern hemisphere, picking can begin in late August through to late October or even November, depending on the mesoclimate and the year's weather. In the southern hemisphere, picking can start in late January and continue through to May.

7.1 Grape ripeness and the timing of picking

The quality-conscious grower aims for physiological ripeness, i.e. tannins and flavour compounds. Also important are matters of sugar, acidity and pH. A lot can be determined by tasting the grape and examining the pip, which, in the case of black grape varieties, should be completely brown when fully ripe. In the weeks and days preceding harvest, the grapes are tested regularly for sugar content. This is usually carried out in the vineyard using a refractometer. The grower and winemaker will rely heavily on their own sense of taste by tasting the grapes in the vineyard. Of course, a technical analysis is also undertaken. The timing of picking is one of the most crucial decisions a grower will make during the vineyard year. This decision is usually made jointly between the grower and the winemaker, unless they are one and the same person!

Wine Production and Quality, Second Edition. Keith Grainger and Hazel Tattersall.
© 2016 John Wiley & Sons, Ltd. Published 2016 by John Wiley & Sons, Ltd.

In cooler climates, the grower may be tempted to delay picking in order to obtain a little extra ripeness, especially of the polyphenols (i.e. flavones, anthocyanins and tannins). Decisions on harvest date also need to be taken with particular regard to the weather forecast. The possibility of autumn rain, which may cause rot, can put pressure on timing. Should there be any existing fungal disease among the grapes, a decision to pick early may have to be made, even though the grapes are still slightly under-ripe.

Where there are different mesoclimates and terroirs within the vineyard, grapes may ripen at different times. Of course, different grape varieties also ripen at different rates. Merlot, for example, ripens earlier than Cabernet Sauvignon – in the Bordeaux region usually some 8–14 days earlier. In hotter climates, ripening can accelerate very quickly, causing acidity levels to fall rapidly, and it is necessary to undertake a speedy picking in order to achieve a well-balanced wine. In hot climates, picking is usually done at night (by machine) or early morning, to enable the grapes to reach the winery while still cool. Also, consideration must be given to the time of day for pickers working in the vineyard, as picking in the midday sun is not beneficial to either grapes or pickers!

7.2 Harvesting methods

There are two basic methods of picking: hand and mechanical. In some European regions, wine laws only allow hand picking. Grapes start to deteriorate from the moment they are picked. The method of picking affects the style and quality of the finished wine. Essentially, the difference is between slow and selective hand picking and the speed of machine harvesting. There are advantages and disadvantages to each method, and the choice may depend on several factors. It can be argued that grapes have a brief window of perfect ripeness, and thus the speed of machine picking is advantageous. Berries can quickly be sunburned, and some varieties, particularly Shiraz, are prone to berry crinkle if not harvested at optimum maturity. A photograph of grapes being harvested by hand at Château Lassegue in Saint-Émilion is shown in Figure 7.1, and a mechanical harvester at work in Romania as Figure 7.2.

7.2.1 Hand picking
This method can be used whatever the training system in the vineyard and on all terrains, however steep or rugged. In many of the world's wine-producing regions, cheap labour is no longer widely available. Labour costs can be substantial, and the team will need housing and feeding. The date of harvest cannot always be accurately predicted well in advance, and so having the skilled labour available when required can sometimes pose problems.

Figure 7.1 Hand picking at Château Lassegue, Bordeaux. Source: Courtesy of Château Lassegue.

Figure 7.2 Mechanical harvester at work in Romania.

Picking by hand is essentially the cutting of part or whole clusters. For some types of wine made by particular vinification techniques, whole bunches are essential. In Champagne, for example, bunches are always pressed whole, i.e. without prior crushing, in order to limit the ingress into the juice of bitter phenolics contained in the skins. Of course, stalks are usually picked along with the grape berries. It is possible to be very selective: damaged parts of the bunch or even individual rotten berries may be removed. In vineyards producing grapes for high-quality sweet wines such as Sauternes in Bordeaux, single berries affected by noble rot can be carefully selected. Obviously, this is very time-consuming and expensive in labour costs.

Gentle handling results in minimal damage to the grapes. Small plastic crates (lug boxes) that hold 12–20 kg of grapes and stack on top of each other without crushing the grapes are now used by many growers. These enable the grapes to be transferred to the winery in good condition. Rapid transport to the winery is important. In very hot areas, such as Rueda in Spain, grapes handpicked in the early morning directly into plastic-lined containers have a covering of carbon dioxide in the form of dry ice put over them to prevent contact with air. The grapes are then chilled as soon as they reach the winery. However, in many regions, it is still common to see crates that hold 50 or 100 kg of grapes, and, when little consideration is given to quality, grapes may simply be dumped straight into the lorry or trailer that will transport them to the winery. Quality-conscious producers will employ a team at a sorting table, either in the vineyard or at the winery reception, in order to eliminate unsuitable grapes and other unwanted materials such as leaves, snails, etc. This material is referred to as *MOGS* – materials other than grapes. There has been considerable technical advancement with sorting tables since the first manual tables were used, as we shall see in Chapters 11 and 25.

7.2.2 Machine picking

In many countries, machine picking has been widely undertaken in the past few decades. The technology of mechanisation has developed machines that can work on a modular basis and can be used for spraying, pruning and picking. There are, however, still many producers who consider even the most advanced machines to handle the grapes roughly and refuse to use them. The capital cost is substantial – it is not just the cost of the machines but also the preparation of the vineyard and the installation of appropriate trellis training systems. Some growers rely on the hiring of a picking machine, but this may not always be available at the required time, especially if the weather in the growing season has been abnormal.

Until recently, the design of mechanical harvesters was such that they could only be used on flat or gently sloping terrain. The berries are usually picked by being vibrated off the vines and are collected in a reception hopper. The vibrating arms can cause damage to vines and even the trellis system. It is important that a precise fruiting zone be controlled on the trellis hedge. Machines can pick a vineyard in a fraction of the time that a hand-picking team needs, which is important where picking time is limited, e.g. bad weather is forecast. Also, mechanical harvesters can operate 24 h a day and have the advantage of being able to undertake picking at night. This is important in hotter regions, ensuring that grapes are picked at cool temperatures and delivered to the winery in good condition. The winemaker does not want to receive hot grapes. A disadvantage of mechanical harvesting is that the process is largely unselective. Most machines in common usage are unable to distinguish between ripe and unripe grapes, or healthy and diseased grapes. However, 'shot' berries, i.e. those that are totally undeveloped, will remain on the vine. A quality-conscious producer will precede the harvesting machine to 'rogue' fruit by picking and dropping to the ground any diseased or damaged berries or clusters. The action of most machines means that only berries can be picked and not whole bunches. So, the stalks, which could act as drainage channels in the early stages of winemaking, or add tannic structure to the wine, will have already been excluded. A serious drawback of machine harvesters is that they do pick MOGs. These can include ladybirds (which can badly taint wine), snails, leaves and vineyard debris.

Figure 7.3 Potassium metabisulfite added to picking bins.

7.3 Style and quality

Ultimately, the method of picking and handling of the grapes, the timing, their condition and the speed of their arrival at the winery will impact on the style and quality of the finished wine. If the grapes are damaged, oxidation will set in, and unwanted skin contact with the juice, particularly for white wines, can lead to excessive phenolics and the loss of aromatics. Sulfur dioxide, in the form of potassium metabisulfite ($K_2S_2O_5$), may be added to the picking bins for protection of the fruit as illustrated in Figure 7.3 and discussed in Chapters 9 and 10. Whatever the method of picking, rapid transport to the winery and speedy processing are critical from a quality perspective.

CHAPTER 8

Vinification and winery design

This chapter looks at vinification, the processes of turning grapes into wine. The processes have become highly developed, particularly in the last 60 years, and are full of scientific complexity. However, it is not the intention of this chapter to explore vinification in detail or engage with detailed science or complicated techniques but to draw the reader into the subject to grasp basic concepts. Winery design and the equipment used for winemaking are also briefly covered.

8.1 Basic principles of vinification

Basically, wine is made from just from grapes, and to produce a 75 cl bottle of wine, the amount of grapes required will be at least 1.1 kg. Needless to say, at harvest time, wineries are a hive of industrial activity. It is important to process grapes as they arrive without delay.

The sugars contained in the pulp of ripe grapes are fructose and glucose, in approximately equal proportions. During fermentation, enzymes from yeast convert the sugars, which are carbohydrates, into ethyl alcohol and carbon dioxide in approximately equal proportions, and heat is liberated:

$$C_6H_{12}O_6 \rightarrow 2CH_3CH_2OH + 2CO_2 + Heat$$

NB: the formula for ethyl alcohol (ethanol) is often abbreviated to C_2H_5OH, or C_2H_6O. Additionally, tiny amounts of other products are formed during the fermentation process, including glycerol, succinic acid, butylene glycol, acetic acid, lactic acid and other alcohols.

Wine Production and Quality, Second Edition. Keith Grainger and Hazel Tattersall.
© 2016 John Wiley & Sons, Ltd. Published 2016 by John Wiley & Sons, Ltd.

The amount of sugar in must and the reducing sugars (glucose and fructose) in fermenting wine, can be determined by using a density hydrometer. A number of different scales can be used to measure density, which can be directly related to sugar content using appropriate formulae or, more readily, prepared tables. Different countries tend to use one or other of the scales available. The scales in most common use are Baumé and Brix, named after their inventers, but Balling, Oechsle and KMW (Klosterneuburg Mostwaage) are used in some countries. 1° Baumé = 1.8° Brix. The scale on a hydrometer is calibrated to give an accurate reading at a specific temperature, often 20°C (68°F), so adjustments to the reading must be made if the liquid is measured at another temperature. When making white wines, every 17 g/l of sugar in the must will, when fermented, will produce approximately 1% abv. For red wines approximately 19 g/l of sugar is needed – red wines will be fermented at higher temperatures, and some of the alcohol produced will evaporate during the winemaking process.

The winemaker has to control the fermentation process, aiming for a wine that is flavoursome, balanced and in the style required. The processes of winemaking are fraught with potential problems, including stuck fermentations (the premature stopping of the fermentation while the wine still contains unfermented sugars), acetic spoilage or oxidation. Throughout the vinification process, it is essential to maintain accurate records, including temperature and gravity readings.

There are key differences in the making of red, white and rosé wines. Grape juice has little colour. For red wine, the colour must be extracted from the skins, and this is usually achieved by fermenting the juice together with the skins at a relatively warm temperature. For white wines, the skins are normally excluded from the process, although some winemakers choose to have a limited skin contact between the juice and skins before fermentation, as this can add a degree of complexity. Of course, white wine can be made from black grapes, as is commonly practised in the Champagne region. Rosé wine is usually made from black grapes whose juice has been in contact with the skins for a limited amount of time, e.g. 4–16 h. During this period, a little colour is leached into the juice, which is then drawn off the skins. The making of rosé wines is discussed in Chapter 14.

8.2 Winery location and design

Wineries vary from centuries-old stone buildings externally of timeless appearance to the modern practical constructions, perhaps built to the design of a specialist winery architect. The winemaker who has inherited the ancestral château with all its low ceilings, narrow doors and dampness may long for the blank canvas available to those with an embryonic project. Conversely,

Figure 8.1 Winery without walls.

when faced with a sterile hangar of his or her newly built winery, a producer may envy the neighbour's ancient property, especially in an age when wine tourism and cellar door sales have become a major contribution to the profits of many businesses. For large producers, efficiency and economy often dictate, and during the past 30 years many large fermentation and bulk storage facilities have been built that have no walls (and sometimes no roof over the vats). Figure 8.1 illustrates a wall-less, industrial winery in Chile.

When planning a new winery, practicalities usually dictate. Keeping the building cool is a major consideration. In a hot climate, the building should be sited on a north–south axis, with the shorter walls into the midday sun. It can also be beneficial not to have windows on the western side, facing the hot afternoon sun. Adequate insulation is required, for the ideal ambient temperature inside is no more than 20°C (68°F). Pumping of must or wine over long distances should be minimised. Some winery designers aim for gravity flow, with grapes arriving or being hoisted to the top level for crushing. The must or wine moves only by gravity to subsequent stages of the winemaking process. Some wineries have been built into hillsides to facilitate the gravity flow principle and minimise the ambient temperature.

Wineries use copious amounts of water, particularly for cleaning purposes. Although water is not an ingredient in winemaking, as a general rule, for every 1 litre of wine produced, 10 litres of water will be used in the winery. Adequate drainage is required throughout, especially in the vat rooms and areas where equipment washing takes place. Ventilation is vital. Every litre of grape juice, when fermented, produces about 40 litres of carbon dioxide. The gas is asphyxiating, and consequently deaths in wineries can, and sadly do, occur.

8.3 Winery equipment

Equipping or re-equipping a winery is very capital-intensive, and some of the items are used for just a few days a year. Equipment utilised includes:
- sorting table(s);
- destemmer(s);
- crusher(s);
- fermentation and storage vats;
- press(es);
- pumps;
- fixed and moveable pipes and hoses;
- filters;
- refrigeration equipment and vat-cooling system;
- barrels, if utilised;
- bottling line, if wine is to be bottled on the property;
- laboratory equipment;
- cleaning equipment.

8.3.1 Fermentation vats

Wine is usually fermented in vats, although barrels are sometimes used, particularly for white wines. Historically, wine was made in either shallow stone tubs called 'lagars', in which the grapes would be trodden with the gentlest of crushing or open wooden vats. Cuboid vats made of concrete (béton) or cement (ciment), became very popular with producers in the early to mid-twentieth century. They can be built in a single or double tier against the walls of the winery, utilising all available space. Until recently, it was regarded as best practice that concrete or cement vats should be lined, to prevent the acids in must and wine attacking the material and to facilitate cleaning. Glazed tiles were often used for this purpose, but they are prone to damage, and wine getting behind them would present a serious hygiene problem. A much better lining material is epoxy resin. In the last few years, concrete vats are having a resurgence in popularity, being prized for, inter alia, their thermal insulation properties. They now come in a variety of shapes, including eggs and amphorae, and are usually unlined, other than a light spraying with cream of tartar. An illustration of concrete eggs at Domaine Jean-Marc Brocard in Chablis is shown in Figure 8.2.

Figure 8.2 Concrete eggs in Chablis.

Stainless steel (inox) vats are now commonplace, but it is interesting to note that it is only from the late 1970s that they gained universal acceptance. When, in the 1960s, stainless steel was installed at Château Latour in Bordeaux, the locals accused the British (who owned the property at the time) of turning the illustrious château into a dairy! At the start of the 1970s, only three Bordeaux properties were using stainless steel, but its presence in winemaking has been commonplace for 30 years. The great advantages of the material are ease of cleaning and the ability to build in cooling systems. There are two grades of steel used for construction: 304 grade is really best suited for red wine fermentations, while the better 316 grade, which contains molybdenum and is harder and more corrosion-resistant, is ideal for both reds and whites – white wines have higher levels of acidity. Stainless steel vats may be open-topped or closed, with a sealable hatch lid and fermentation lock. The stainless-steel vats in the new fermentation room at Château Pichon-Lalande in Pauillac, Bordeaux are illustrated in Figure 8.3.

Variable-capacity vats are also available and are particularly useful when only partly filled. These have a floating metal lid, held in place by an inflatable plastic tyre at the perimeter. Fibreglass vats are still occasionally used as a less costly alternative to stainless steel.

Although, in the late twentieth century, many producers replaced their wooden (and concrete) vats with stainless steel, in recent years wood has

Figure 8.3 Vats at Château Pichon-Lalande.

Figure 8.4 Wooden vats at Château Pontet-Canet.

regained popularity as a material for vat construction at some wineries. The difficulties of maintaining and sanitising the vats are ever present, but there is less risk of the wine being affected by reductivity, a topic discussed in Chapter 21. Wooden vats at Château Pontet-Canet in Pauillac, Bordeaux are illustrated in Figure 8.4. Equipment other than vats used in the winery will be discussed in subsequent chapters.

CHAPTER 9

Red winemaking

This chapter describes the processing of grapes to make red wine. Various approaches can be taken in the production of red wine according to the style, quality and quantity of wine being made. In this chapter, we consider the processes of destemming and crushing, must preparation, fermentation and extraction, pressing, malolactic fermentation, maturation and blending. It should, however, be borne in mind that the matters discussed here are neither prescriptive nor exhaustive.

9.1 Sorting, destemming and crushing

The grapes may be sorted on arrival at the winery as discussed in Chapters 7, 11 and 25. The grape stalks are usually removed to prevent bitterness tainting the juice. Following this, the grapes can be lightly crushed. These tasks can be performed by the same machine, a destemmer–crusher, or the destemmer and crusher may be separate machines. At the destemming stage, the grapes are fed via a hopper into a slotted cylinder, which contains a rotating rod fitted with 'propeller-like blades'. As this rotates, the berries pass through the slots, leaving the stalks behind – these are then expelled from the machine and can be used for vineyard fertiliser. The destemmed grapes may be sorted before crushing to remove rotten or under-ripe fruit. A destemmer, situated over a sorting table, is illustrated in Figure 9.1. At the crushing stage, the grape berries are passed through a set of rollers, which can be adjusted to give the chosen pressure in order to release the juice. The juice and skins (must) are then transferred, usually by pump, into fermentation vats. The amount of must obtained from each tonne of grapes will vary according to, inter alia, the

Wine Production and Quality, Second Edition. Keith Grainger and Hazel Tattersall.
© 2016 John Wiley & Sons, Ltd. Published 2016 by John Wiley & Sons, Ltd.

Figure 9.1 Destemmer placed over sorting table.

grape variety, berry size and ripeness. The figure will usually be in the region of 650–730 litres, but with a quality-conscious winemaker, processing small berries, it can be as little as 550 litres.

For certain wines, where whole bunches are required for fermentation, e.g. Beaujolais, the grapes will not be destemmed or crushed. There is a growing trend among some red winemakers to include at least a percentage of whole grapes in their fermentations, including those destined for classic, full-bodied wines.

9.2 Must analysis

The winemaker needs to know the composition of the must in order to decide on necessary adjustments and make decisions as to the precise path of winemaking. The following analyses are regarded as essential:

- temperature;
- must density;

- pH;
- total acidity, and ideally tartaric and malic acid components;
- free and total sulfur dioxide (see Section 9.3.1).

9.3 Must preparation

The must, which comprises grape juice with seeds, skins and pulp, now has to be prepared for fermentation. Various additions and adjustments may be undertaken.

9.3.1 Sulfur dioxide (SO$_2$)

This is the winemaker's universal antioxidant and disinfectant, and is used at many stages in winemaking. It may be used as a gas or in the form of potassium metabisulfite. However, only a percentage of the sulfur dioxide in the wine, the 'free' SO$_2$, acts in a protective manner. The other proportion becomes 'bound' with the wine and is inactive. To prevent fermentation starting prematurely, it may be added at crushing and/or at initial must preparation to inhibit the action of wild yeasts and bacteria. These organisms, naturally found on grape skins, require oxygen for growth. Wild yeasts, which can cause off-flavours, die when 4% alcohol is reached, and many winemakers inhibit any such fermentation. Naturally occurring wine yeasts, of the *Saccharomyces* genus, found in the vineyard and the winery, work without the action of oxygen and thus can work even if the must is treated or blanketed with sulfur dioxide. In many parts of the world, winemakers now prefer to use selected cultured yeasts for greater control and reliability, and for specific flavours. It should be noted that producers often speak of wild yeast fermentation, when referring to use of the natural yeasts that come into the winery on the grapes or are present on surfaces in the winery.

9.3.2 Must enrichment (chaptalisation)

In cooler climates, grapes often do not contain enough sugars to produce a balanced wine. This problem may be addressed by the addition of sucrose to the must in the early stages of fermentation, a process known as chaptalisation – Jean-Antoine Chaptal was a French Minister of Agriculture who first authorised the process in 1801 as a way of disposing of surplus sugar. It is important that only the minimum necessary amount be added, or further imbalance will be created. Must enrichment is not permitted in many hot countries. The EU vineyard area is divided into zones according to crude climatic conditions, and the amount, if any, of chaptalisation allowed varies according to the zone. In some regions and countries, rectified concentrated grape must is used instead of sugar.

9.3.3 Acidification

This may be necessary if the pH of the must is too high, that is, if the acidity is too low. For red musts, a pH of 3.8 might be regarded as a maximum, but this varies according to region of production and the wine style required. The addition of tartaric acid is the usual method of acidification. The addition of malic acid is not permitted within the EU, although it is not uncommon in Argentina and some other New World countries. Acidification is generally not permitted in cooler regions of the EU (classified as regions A and B), but derogations are possible. For example, in 2003, Regulation (EC) No. 1687/2003 permitted acidification on account of the exceptional (hot) weather conditions. Figures indicate that the worldwide average addition of tartaric acid in winemaking equates to 1.6 g/l, so bearing in mind it is not used in many regions and countries, the average addition is much greater than this.

9.3.4 De-acidification

This may be necessary if the pH of the must is too low, perhaps below pH 3.2, again depending on region and required wine style. It is not permitted in warmer regions of the EU. There are a number of materials that may be used, including calcium carbonate ($CaCO_3$), perhaps better known as chalk, potassium bicarbonate ($KHCO_3$) and potassium carbonate (K_2CO_3). Another agent that may be utilised is Acidex®, which is a double-salt seeded calcium carbonate designed to reduce both tartaric and malic acids in must or wine. The product was developed in Germany, where the cool climate often produces grapes of high acidity. NB: calcium carbonate only reduces the tartaric acid content.

9.3.5 Yeast

Cultured yeasts may be added, or the winemaker may simply utilise the natural yeasts present on the skins and in the winery. One reason for using cultured yeasts is to ensure a complete fermentation, particularly of high-density musts. Each strain of yeast has its own flavour profile, and, particularly in high-production wineries, the winemaker requires reliability and the control that these facilitate. Conversely, other producers regard natural yeasts as an extension of 'terroir'.

9.3.6 Yeast nutrients

As living organisms, yeasts need nutrients. The following may be added to the must.

9.3.6.1 Diammonium phosphate [$(NH_4)_2HPO_4$]

Diammonium phosphate, often referred to as DAP, is usually added at a rate of 200 mg per litre of must, to help ensure that all the sugars are totally fermented and to stop the formation of hydrogen sulfide (H_2S), which is most undesirable. Its use is common in New World countries, particularly when the musts are nitrogen-deficient, which may lead to the production of H_2S.

9.3.6.2 Thiamine (vitamin B₁)

Sometimes, together with other B group vitamins, thiamine may be added in the early stages of fermentation to help boost yeast populations and prolong their life. However, the yeast *Brettanomyces*, regarded by most as a spoilage yeast in wine production and therefore generally undesirable, needs thiamine to grow, so additions need to be undertaken with caution.

9.3.7 Tannin

This powder may be added to red musts when the winemaker is seeking additional tannic backbone. It also helps to 'fix' the colour of red wines.

9.4 Fermentation, temperature control and extraction

9.4.1 Fermentation

The fermentation of red wine takes place with grape solids present, in order to extract colour, tannins and flavours from the skins. Initially, the fermentation can be very tumultuous, so head space must be allowed when filling the vats, but as more sugar is converted, the rate slows down. In the majority of cases, the fermentation is continued until the wine is dry or off-dry, and, depending upon the density of the must, the final alcohol concentration is generally 11–14.5% by volume.

9.4.2 Temperature control

The fermentation process is a turbulent one and creates heat naturally. During red winemaking, fermentation may begin at about 20°C, but temperatures may rise to 30–32°C. Yeasts cease to work if the temperature rises above approximately 35°C. Some form of temperature control may be necessary to prevent this happening before the sugars are fully fermented and to maximise aromas and flavours. It is only in the past few decades that winemakers have had the equipment and ability to be able to have effective control of fermentation temperatures.

One hundred grams of sugar per litre of juice has, when fermented, the theoretical potential to heat the juice by 13°C. If grape must of 11.1° Baumé (which equates very approximately to a potential of 11.1% abv) commences fermentation at 15°C, the theoretical increase in temperature could be to 41°C. If the fermentation of must with a Baumé of 14.5° (potential alcohol of very approximately 14.3%) commences at 20°C, the theoretical temperature increase is to 54°C. Of course, some of this heat is naturally dissipated during the period of fermentation through the walls of the vat and, with the rising carbon dioxide (CO_2), through the juice to the surface. Larger vats have a lower percentage of surface area so will dissipate less heat.

Good extraction of tannins requires warm fermentations. However, cooler fermentations aid the growth of yeast colonies and can yield higher alcoholic degrees after fermentation to dryness. Cooler fermentations aid the development of aromatics. The warmer the temperature, the less time the fermentation takes. Accordingly, managing the temperature can be quite a complex exercise. A winemaker may decide to start a vat fairly cool, at say 20°C, and allow it to rise naturally to around 30°C to aid extraction. In the latter stages, the vat may be cooled to 25°C or so, to ensure complete fermentation to dryness. In the cool underground cellars of regions like Burgundy, the temperature of small vats or barrels can be self-regulating. Cooling equipment may be required for larger vats. Wine can be pumped through heat exchangers to reduce (or increase) temperature. Stainless steel tanks are now commonly wrapped with water or glycol cooling jackets. In concrete or wooden vats, a metal cooling coil or plate (drapeau) can be utilised, and the latest concrete vats can have a cooling system built into the walls.

9.4.3 Extraction

The usual process for red wines is for the grape mass to be fermented in vertical vats, either open-topped or closed. The solids and skins rise to the surface with the CO_2 and create a floating cap, as shown in Figure 9.2. This is a disadvantage

Figure 9.2 Cap of grape skins in vat.

Figure 9.3 Looking down on remontage.

because the skins need to be in contact with the juice for there to be good extraction of colour and tannins. Also, acetic bacteria thrive in such a warm, moist environment, risking spoilage of the juice. Consequently, during the process, the juice is drawn out from a valve near the bottom of the vat and pumped up and sprayed over the cap to soak it. This process, known as pumping over or remontage, has the additional benefit of aerating the must, which helps to boost the yeast colonies. The process may take place several times a day, particularly in the early stages of fermentation. An illustration of remontage is shown in Figure 9.3. An alternative technique of pigeage, simply punching down the cap, is used for some varieties. Traditionally used for Pinot Noir which needs a very gentle extraction process, it is now gaining popularity for red winemaking generally. Although vats can now be fitted with mechanical pigeage equipment, in many wineries submerging the cap is done by hand with wooden paddles, sticks or even the feet of the cellar staff precariously suspended over the fermentation vats.

9.4.4 Fermentation monitoring
The winemaker will regularly monitor the fermentation, analysing the juice and recording the results. The gravity and temperature will be checked daily, indeed several times a day during the early period of ferment. Well-equipped wineries

may have a programmable computer control system to adjust and keep the fermentation temperatures within the desired parameters. Laboratory analysis of juice may also be undertaken. Process sensor technology is the latest aid being introduced whereby the juice is being continuously analysed and data captured.

9.5 Maceration

Depending on the style required, the wine may be left to soak with the skins after completion of the alcoholic fermentation, until sufficient flavour and tannins are extracted. This maceration could vary from 2 or 3 days up to 28 days. If an 'early drinking' red is required, the juice may be drained off the skins at some point during the fermentation. Most, if not all, of the colour will have been extracted after 4 days of fermentation on the skins. Alternatively, if the winemaker believes that a post-fermentation maceration would not be beneficial (high alcohol levels may extract unwanted hard tannins), the tank may be drained hot, i.e. immediately after the alcoholic fermentation, before the wine has cooled.

9.6 Racking

Racking is the process of transferring juice or wine from one vessel to another, leaving any sediment behind. After the fermentation and any maceration, the liquid contents of the fermentation tank will be run off to another vat. This juice is called free-run juice. The skins and other solids are left behind in the fermentation vat. These will be transferred to the press to obtain further juice. Racking will also take place at various other times in the winemaking/maturation process, to remove the wine from lees and sediment, and clarify it. Aeration can also take place during the racking process, and SO_2 may be added if necessary.

9.7 Pressing

The juice released from the press will naturally be higher in tannin and colouring pigments. Between 10 and 15% of the total juice comes from the pressing process. Some producers choose to make their wine, or their top-quality wine (sometimes referred to as 'grand vin') only from free-run juice. More than one pressing may be carried out, but with each pressing the press wine becomes coarser. Presses are controlled to exert varying levels of pressure at different stages, often gentle at the beginning but harder with each subsequent pressing. A variety of types of press are available, including basket presses, horizontal plate presses and pneumatic presses, and the characteristics of these will be considered in Chapter 11.

9.8 Malolactic fermentation

This usually follows the alcoholic fermentation and so is sometimes referred to as the secondary fermentation. Yeast is not involved. In fact, it is a transformation caused by the action of strains of bacteria of the genera *Lactobacillus*, *Leuconostoc* and *Pediococcus*. Harsh malic acid (as found in apples) is converted into softer-tasting lactic acid (as found in milk). The process is almost always undertaken for red wines. One gram of malic acid produces 0.67 g of lactic acid and 0.33 g of CO_2. Malolactic fermentation (MLF) can be induced by warming the vats, or inoculating with strains of lactic acid bacteria. The ideal conditions for the fermentation to take place include a temperature of approximately 20°C and SO_2 levels close to zero. Alternatively, it could be blocked by treating the wine with SO_2 and/or keeping the wine cool, although this would usually only happen in the making of white wines. More and more winemakers now undertake the MLF at the same time as the alcoholic fermentation, which helps avoid microbial alteration and enables the wine to be given protective additions of SO_2 immediately after completion of the ferments.

9.9 Blending

This is an important operation in wine production. Once fermentation has finished there will be several vats or barrels containing wines from different vineyards, blocks within vineyards and varying ages of vine. Individual grape varieties, having been picked according to ripeness, will have been fermented separately. Blending of these various vats will then take place in order to achieve the desired final style and quality of wine. The prime reasons for blending are to have a product that is greater than the sum of its parts, to even out inconsistencies and perhaps to maintain a brand style. This selection process could involve a great number of component parts. Where different grape varieties are blended together, the winemaker considers what proportion of each variety is included in the blend. Some varieties or vineyard blocks may perform better in a particular season. Where high quality is paramount, some vats may well be rejected and used in a lower quality wine or sold off in bulk.

9.10 Maturation

Immediately after fermentation, wines may taste harsh and fairly unpleasant. A period of maturation is usually required, during which chemical changes take place including softening of tannins and integration of acidity. However, some wines that are intended for early drinking, such as inexpensive or branded wines, will have very little maturation, having been made with minimal extraction of tannins. The choice of maturation vessel and the period of time depend upon the style of wine being produced, quality and cost factors. There are many types of maturation vessels, including vats made of stainless steel or concrete and wooden barrels. Of course, stainless steel is impermeable to gases such as oxygen, and stainless steel vats are ideally suited to temperature control. They may be used when long-term, oxygen-free storage is required, such as when inexpensive wines are held prior to being 'bottled to order'.

Most high-quality red wines undergo a period of barrel maturation – usually somewhere between 9 and 22 months. During the time in the barrels, the wine will undergo a controlled oxygenation and absorb some oak products, including wood tannins and vanillin. When the barrels have been filled, they will be tapped with a mallet to dislodge air bubbles, which rise to the surface of the wine, thus removing oxygen. During the period in the barrel, the wine may be racked several times to aid clarification. Red Bordeaux wines, for example, are traditionally racked four times during the first year's maturation, and perhaps once or twice in the second. However, many producers, particularly when advised by winemaking consultants, have reduced the number of rackings and moved to minimal handling of the wine. Barrels are regularly topped up with a similar wine to replace wine lost to evaporation, although there is also a view that if the barrels are securely sealed the ullage at the top is a partial vacuum, and topping up may not be beneficial. The topic of barrel maturation is further considered in Chapter 12, and the processes of preparing the wine for bottling, are detailed in Chapter 13.

CHAPTER 10

Dry white winemaking

In this chapter, we consider the ways in which the techniques used to make white wines differ in some key areas from those used for reds. We introduce concepts unique to white wine production, such as lees ageing. The sequence of operations for white wine production may be as follows.

10.1 Crushing and pressing

10.1.1 Crushing

On arrival at the winery, white grapes should be processed with minimal delay to avoid deterioration and the onset of premature fermentation. In hot climates, dry ice, the solid (frozen) form of carbon dioxide (CO_2), may be placed on the bins of grapes. This releases CO_2 gas, which will help protect the grapes from bacterial spoilage and oxidation of any damaged fruit. If the fruit has arrived at the winery hot, the grapes should be chilled before the crushing stage, perhaps in a blast chiller room. In most cases, they will be de-stemmed and lightly crushed before pressing. It is becoming increasingly popular for the winemaker to allow some hours or even days of skin contact with the juice at low temperatures before the grapes go to the press. This may extract aroma compounds and enhance flavours, but care must be taken not to extract bitter phenolics.

When white wine is being made from black grapes, crushing does not take place, or colour would be leached into the juice. Instead, whole clusters are sent to the press. Many winemakers choose to do this with clusters of white grapes, too, believing this gives a greater purity of juice.

Wine Production and Quality, Second Edition. Keith Grainger and Hazel Tattersall.
© 2016 John Wiley & Sons, Ltd. Published 2016 by John Wiley & Sons, Ltd.

10.1.2 Pressing

Unlike in the red wine process, separation of juice and skins occurs before fermentation. Enzymes to help juice extraction may be added, and pectolytic enzymes utilised to prevent too much frothing during the fermentation and possible haze in the finished wine. Of course, gentle pressing, or alternatively allowing crushed grapes to drain in a vertical drainer, results in better-quality juice. It is important that the pips remain intact in order to avoid the release of bitter oils. The skins of white grapes are not used during the fermentation process.

10.2 Must preparation

The juice is drained from the press or drainer into settling tanks. If the must is not cleared of solid matters, off-tastes can result. A simple settling may take place over a period of 12–24 h, a process known in France as débourbage (bourbs are the suspended solids). The must is then racked to another tank for fermentation. Clarification can also be achieved by using a centrifuge, which speeds up the process, or by filtration. Protein may be removed by treating the must with bentonite. This is a form of clay earth that acts as flocculent, attracting and binding fine particles, which then settle out of suspension. However, bentonite can also remove desirable aromas and flavours. It is important to carry out bench tests to determine the amount required, which might be as little as 6 g/hl or as much as 180 g/hl. If the must is over-clarified, fermentation may proceed slowly, and the wine may not ferment to completion. Yeast nutrients attach to solid matters within the juice, making them accessible to yeasts, and removal of these solids has a detrimental effect on yeast growth. The must may be passed through a heat exchanger to lower the temperature. This not only prevents a premature onset of fermentation but also preserves freshness and flavours. The must is treated with SO_2 to prevent aerobic yeasts and spoilage bacteria. Must enrichment (chaptalisation) may take place if necessary and legal.

10.3 Fermentation

The must is pumped directly to the fermentation vats or barrels. With careful winery design, this process can be achieved simply by gravity, for pumping is by nature a harsh process. Selected cultured yeasts may be added, although the use of so-called 'wild yeasts', i.e. natural yeasts present on the grape skins and in the winery, may be used to enhance particular characters. White wine is usually fermented cooler than red, between 10 and 20°C to preserve primary fruit flavours. Of course, cooler fermentations take longer but result in the retention of more aromatics, which are not lost to the atmosphere. Each vat is under temperature control, usually with its own chilling system. While cool fermentations are desirable for aromatic whites, less cool fermentations are used when full-bodied wines are desired. Some white wines are fermented in barrels to give particular characteristics to the wines and a harmonious integration of oak flavours. Barrels also give a regular supply of oxygen, required by the yeast. The choice of new or used oak from particular origins is made. It may be that different lots are fermented in new, second and third fill oak barrels (which give a more restrained oak influence). These will be blended together at a later stage, perhaps also with an unoaked, vat fermented component.

10.4 MLF

MLF may follow alcoholic fermentation to soften acidity and create a 'rounder' texture. Some white grape varieties, e.g. Chardonnay, are particularly suited to MLF. MLF gives the wine a slight buttery nose, a certain amount of complexity, and a creamy texture. Other varieties that are valued for their crisp acidity, such as Riesling, do not usually undergo the process. Also, in some countries, the crispness of wines that have not undergone the malolactic fermentation is a desirable feature. For example, the lively acidity of a Sancerre, made with the Sauvignon Blanc variety in the Loire Valley in France, is an essential part of its character. Following the MLF, the wine is racked into clean vats or barrels and SO_2 added.

10.5 Lees ageing

After fermentation, the wine may be left on the lees, which, from time to time, may be stirred. This stirring operation is known as bâtonnage and adds yeasty, and perhaps creamy, flavour to the wine. NB: the process is not suitable for red wines – as the yeast cells break down, they release, inter alia, proteins that bind with tannins. This helps give a softer mouthfeel to white wines but would reduce the structure and ageing potential of reds. The lees also scavenge oxygen, and the wines thus aged will require less sulfur additions. As an alternative to lees stirring, the barrels may be rolled. This process may by simplified when the barrels are stacked on racks fitted with rollers, e.g. the Oxoline system.

Vat-matured wines may also be left on the lees. When considering lees ageing, we should distinguish between the gross lees, the large lees from the fermentation, pre-racking, and the fine lees that will fall after the wine has been racked for the first time. Ageing on lees may take place on either gross or fine lees, depending on the textures sought, but care must be taken to avoid reduction, as the lees will take out oxygen. Stirring of lees may also take place in vat: a propeller type of device may be inserted through the racking valve at the side of the vat to rouse the settled sediment. It should be noted that some varieties, e.g. Chardonnay, may benefit from lees stirring, while others, e.g. Riesling, will suffer if the lees are roused.

10.6 Maturation

Many white wines are stored in stainless steel or concrete vats until ready for bottling. It is important that oxygen be excluded, and the vats should be kept either completely full or blanketed with N_2 or CO_2. White wines fermented in barrel may undergo subsequent barrel maturation, and tank-fermented wines may also spend some months in barrels. Barrel and other forms of oak maturation will be discussed in Chapter 12.

CHAPTER 11

Red and white winemaking – detailed processes

So far, we have studied the basics of red and white winemaking. In this chapter, we will consider in more detail some of the decisions to be made and processes that may be undertaken, including the different methods and equipment that may be used. We will concentrate in this chapter on the de-stemming and crushing decisions, methods of must concentration, choice of yeasts, colour and flavour extraction, and choice of press and methods of pressing. Clarification, including fining, filtration and stabilisation, will be considered in Chapter 13.

11.1 Must concentration

There are several methods that may be used to extract colour, concentrate flavours and achieve tannin balance.

11.1.1 Must concentrators and reverse osmosis

Grapes may be received that are wet from rainwater or that have not reached complete ripeness. The inclusion of rainwater at crushing can cause dilution of juice, solids (sugars and other components). When under-ripe grapes are crushed, the sugar level may be lower than required to make a good wine with the right level of alcohol and the required balance. In both instances, the solution may be to remove water.

Must concentrators use vacuum evaporation to remove water, thereby increasing the percentage of sugars, colour and flavour components, and tannins in the must. A must concentrator manufactured by REDA is shown in Figure 11.1. An alternative technique is to use reverse osmosis (RO). This is a membrane separation process in which a microporous membrane acts as a molecular sieve,

---

Wine Production and Quality, Second Edition. Keith Grainger and Hazel Tattersall.
© 2016 John Wiley & Sons, Ltd. Published 2016 by John Wiley & Sons, Ltd.

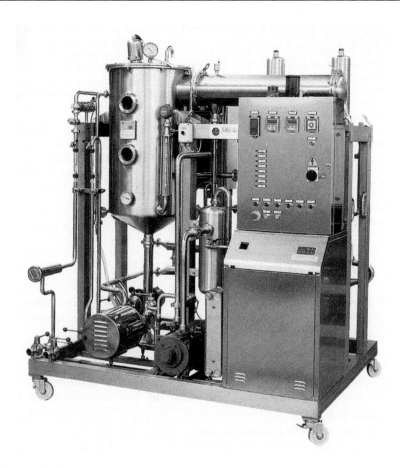

Figure 11.1 Must concentrator.

allowing water molecules to pass through, but trapping larger molecules (sugars, etc.), which are then concentrated. Usually, only a part of the must is put through the process, and the concentrated must is then blended with the unprocessed portion. RO can yield must capable of creating full-bodied, complex and high-quality wines. However, it should be noted that by using these methods, all the remaining components are concentrated, some of which (e.g. green tannins) might not be beneficial to quality. There will be an increase in total, tartaric and malic acids, so the winemaker must take care to maintain balance in the must and resulting wine. RO for the purpose of concentration of musts to be fermented into wine was only permitted in the EU from 1999. The relevant EU legislation was Council Regulation (EC) No. 1493/1999 (EU, 1999), and since superseded by Council Regulations No. 479/2008 (EU, 2008) and Council Regulation (EC) No. 491/2009 (EU, 2009a), Commission Regulation No. 606/2009 (EU, 2009b).

The maxima set by the EU Regulations and International Organisation of Vine and Wine (OIV) in the International Code of Oenological Practices (OIV, 2015a, 2015b) for must concentration by RO are a maximum 20% decrease in must volume and a maximum 2% increase in potential alcohol.

RO can also be used to reduce the alcohol level in over-concentrated fermented wine, as discussed below.

11.1.2 Cryoextraction

Cryoextraction is an alternative method of must concentration but is seldom used, other than for the making of so-called *'icewine'* in Canada. Grapes are frozen at temperatures between −5°C and −10°C, and then crushed and pressed. Water in the grapes is turned to ice, leaving sugars and other components concentrated.

11.2 Methods of extraction

Crucial to the quality and style of red wines is the extraction of colour and flavour-giving compounds from the grape skins. Some of these have already been mentioned, but we will now consider them further.

11.2.1 Cold soaking (pre-fermentation maceration)

Some producers hold the must at a low temperature and carry out a pre-fermentation maceration on the grape skins. This method was introduced in the Burgundy region of France by Guy Accad, an oenologist who, from the late 1970s to the 1990s had much influence upon the winemaking at many top estates. Accad showed that soaking the skins in the pre-fermentation (watery) juice results in fresher, cleaner wines with greater finesse than when the skins are macerated in the post-fermentation (alcoholic) liquid. The temperature of the cold soak can vary from 4°C to 14°C, and the duration may be as short as 24 h or as long as 8 days. Cold soaking does pose some microbiological risks. It is not appropriate if the grapes are damaged or diseased, and many winemakers prefer to start the alcoholic fermentation immediately after crushing.

11.2.2 Pump overs – remontage

Remontage (pumping over) is very widely practised. In its simplest form, fermenting juice is drawn from near the bottom of the vat and pumped to the top, where it is sprayed over the cap of skins, pushed to the surface of the wine by the rising carbon dioxide (CO_2) gas. The operation may be carried out manually, including physically moving the hose to spray and soak all the skins. Closed vats may be fitted with timed automatic systems that pump the juice up a fixed pipe at the side of the vat; the juice then falls onto a rotating diffuser, which gives a more gentle spray onto the cap. When additional aeration is

Figure 11.2 Aerated pump over.

required, as the juice is drawn from the vat it is allowed to splash into a large vessel from where it will be pumped up, as shown in Figure 11.2. The number, timing and method of pump overs will impact upon the style and quality of the finished wine. Many producers regard pump overs as the best method of extraction during fermentation – for example, it is the only method used at the very highly regarded Bordeaux *cru classé* Château Pontet-Canet.

11.2.3 Rack and return (délestage)

One procedure now widely undertaken is rack and return, also known as déle-stage. One of the problems with pumping over is that there can be a channelling effect – the descending juice forms channels in the cap that has risen to the top of the fermenting wine and flows down through these, beating out hard tannins from the adjacent skins. With the rack-and-return technique, the juice content of the vat is racked to another vat, leaving behind the cap of grape skins, which loosens and falls to the bottom. The juice that had been situated directly under the cap is high in polyphenols, and these are now dispersed. To give an aeration, which will help boost yeast colonies, lessen the risk of reductive aromas and polymerise long-chain tannins, the juice from the vat being drained may be splashed into a container prior to pumping into the new vat. Seeds, or at least some of them, may be removed at this time, in order to lessen the risk of bitterness being extracted. The fermenting juice is then returned to the original vat, and the cap rises to the top, leaching much colour in the process. It should be noted that the total tannins extracted using délestage may be greater than with pump overs, the advantage being that tannins that are less hard are leached. Most winemakers feel that délestage should not be undertaken at a late stage in the fermentation process, as this could result in greater astringency.

11.2.4 Punching down – pigeage

Punching down, which is perhaps the oldest of all the methods of extraction, is regarded as a more delicate method than pumping over or délestage. When undertaken manually, this is very hard work – ideally, the cap of skins should be punched down to the base of the tank, and the operation undertaken perhaps twice a day. Recently, mechanical pigeurs have been introduced that can be moved over the tanks as required. Pigeage damages the grape skins less than if they were passed through a pump, as would happen with the remontage and délestage processes.

11.2.5 Rotary vinifiers

These are large, horizontal, cylindrical vessels that rotate during the fermentation process, thereby tumbling the grapes and promoting contact between the skins and juice. They eliminate the need for punching down or pumping over and quicken maceration. Such processing is particularly suitable when the aim is a soft, easy-drinking red of medium body.

11.2.6 Thermo-vinification – heat extraction

The term thermo-vinification is often used for the process of extracting colour for red wines by heating crushed grapes for 20 or 30 min to a temperature of 60–82°C. This breaks down the vacuoles of the skin cells that contain anthocyanins and tannins, rapidly releasing the pigments into the juice. After pressing and cooling of the juice, the fermentation takes place off the skins, in the manner of a white wine. Wines made using this method often have a vibrant purple colour. They can sometimes taste rather cooked and lack definition, so the process is not used for fine wines.

11.2.7 Flash détente

Flash détente is a method to maximise extraction of colour and tannins from skins but minimise the extraction of unwanted components. These include pyrazines, particularly methoxypyrazines, which give wine a green taste, and the aromas and flavours of mould or rot, if the grapes are infected. The equipment is expensive and is therefore most likely to be found in 'custom crush' wineries that process fruit from numerous growers. Grapes are rapidly heated to over 80°C and transferred to a vacuum chamber where they are quickly cooled. The vacuum weakens the grapes, and the skin cells burst from the inside, allowing the extraction of anthocyanins and soft tannins in the pre-fermentation (watery) juice. Some steam is released, carrying with it pyrazines and unwanted volatiles, leaving the remaining juice slightly concentrated. The steam is condensed, and if desired, a percentage may be added back to the must. The resulting wines are not destined for ageing, and the system is not used for fine wines.

11.2.8 Whole grape fermentation, carbonic and semi-carbonic maceration

Most Beaujolais and many other light-bodied red wines are made by the distinctive method of whole grape fermentation. Whole bunches of grapes are tipped into a vat that has been sparged with CO_2. To some extent, the grapes in the bottom of the vat are pressed by the weight of those above. Within the berries that remain whole, an enzyme is secreted, resulting in a fermentation, up to approximately 3% alcohol, without the action of yeasts. This contributes to the jammy style of the wines. Thereafter, the yeast enzymes take over to complete the fermentation. The fermenting wine is allowed to macerate or soak on the grapes for 4–10 days, depending on the style of wine required. The activities taking place in the different levels of the vat are very complex.

11.2.9 Fixing colour

Steps may be taken to ensure that a red wine will hold its colour. The use of wooden staves inside a vat of wine is one way of achieving this, and the tannins released by the wood bind the colour to the wine. The process is often used in New World countries. Figure 11.3 shows inner staves inside a concrete vat.

11.2.10 Post-fermentation maceration

Historically, after fermentation, red wine often had a very lengthy period macerating on the skins, before the free-run juice was run off and the skins pressed. During this time, tannins would be absorbed, which would help to fix the colour and give the wine structure, but the resulting wine would often be very firm in the mouth. During the past few decades, most producers have shortened this period, especially for wines that are not destined for lengthy bottle ageing. In order to produce a soft style, some producers drain the vat 'hot', i.e. before the fermentation has finished.

11.3 Macro-, micro- and hyper-oxygenation

As we have seen, oxygen can be both friend and enemy to wine. During fermentation, there is little chance of the wine oxidising, and indeed oxygen is necessary to maintain healthy yeast colonies. After fermentation, wines may

Figure 11.3 Inner staves in vat.

benefit from absorbing tiny amounts of oxygen, for this will help polymerise the polyphenols and help fix the colour of the wine. There are three methods by which must or wine can be deliberately oxygenated, depending upon the purpose.

11.3.1 Hyper-oxygenation
This is adding large amounts of oxygen to white must pre-fermentation. Flavones and tannins are oxidised, becoming brown insoluble polymers that are removed in the racking following fermentation. The benefits include a reduction in bitter tastes, volatile acidity and acetaldehyde, and a longer shelf-life for the finished wine

11.3.2 Macro-oxygenation
This is adding large amounts of oxygen during the fermentation. It may be undertaken as 'rack and splash' in which the fermenting wine is aerated when returned to the vat or can be incorporated during pump overs. Alternatively, a machine known as a 'cliquer' may be used, particularly in the final stages of fermentation. Macro-oxygenation can also be used to try and restart stuck fermentations, but it is commonly undertaken to help soften harsh tannins and to help to ensure fermentation to dryness.

11.3.3 Micro-oxygenation

Micro-oxygenation (MOX) is the continuous addition of tiny amounts of oxygen to wine in order to improve aromas, structure and texture. It may be carried out before or after the malolactic fermentation, with the former perhaps giving the greatest benefits. The process replicates the absorption of oxygen through the wood that occurs naturally when wine is matured in barrels. The process must be carefully controlled. It is essential to know the amount of dissolved oxygen in the wine at the outset, and the rate and amount of the addition (usually a total of 10 ml/l), for there should be no accumulation of dissolved oxygen in the wine during the process. MOX can reduce undesirable herbaceous or sulfur characters. Its proponents claim that it helps stabilise colour, brings out fruitiness, softens and enriches tannins, and reduces the risk of the wine producing sulfides.

11.4 Removal of excess alcohol

RO is one relatively inexpensive method of removing excess alcohol from wines. An alternative method is to draw off a proportion of an over alcoholic wine and send this to a specialist facility that utilises a spinning cone column (SCC). This is a distillation column through which wine is passed twice. The first distillation extracts the volatile aromas, and the second removes the alcohol. The aromas are then returned to the 'wine', which is sent back to the winery and blended back into the bulk. *ConeTech* at Santa Rosa, California, operate the world's largest SCC plant.

11.5 The choice of natural or cultured yeasts

A decision is made whether to use the natural yeasts present on the grape skins and predominantly in the winery building (sometimes referred to as indigenous or native yeasts) or to inhibit these and work with cultured yeasts. Cultured yeasts may comprise a single or multiple strains of *Saccharomyces*, usually *Saccharomyces cerevisiae*, although *Saccharomyces bayanus* (which is tolerant to high alcohol levels) may be used. Each yeast strain has a flavour impact. To use a cultured yeast, which is generally purchased in dehydrated, powder form, the powder is first rehydrated with water at approximately 38°C. Must from the vat to be fermented will then be added, and this begins to ferment. Before being added to the vat, the yeast mixture must be brought to a temperature that is no more than 10°C above that of the must in the vat; otherwise the thermic shock would affect yeast viability. Yeast booster machines, such as that manufactured by Kreyer, can simplify the preparation of the yeast starter and adjust the temperature to the same

Figure 11.4 Yeast booster.

as the must in the tank. An illustration of a Kreyer yeast booster is shown in Figure 11.4.

Many Old World producers favour the use of natural yeasts, particularly for red wines, believing that they are a part of the 'terroir' of the vineyard, and contribute to the individual character of the wines. Others, particularly those producing on a large scale, prefer the consistency of their chosen cultured yeast strains. In the New World, most producers consider natural yeasts far too risky to use. Control is the aim, and cultured yeasts can be selected for their individual characteristics, including flavours, ability to perform at various temperatures and formation of lees. Natural yeasts may, on occasions, be instrumental in producing higher-than-desired levels of volatile acidity, together with acetaldehyde and hydrogen sulfide.

11.6 Destemming

Some winemakers believe that stems can perform a valuable function. For example, in the Rhône valley, stems are often retained in the vinification process. They help keep the cap of skins on top of the fermenting vat less compact, add firm tannins to the wine and suppress any tendency for overt jammy fruitiness. If stems are included, it is important that they remain reasonably intact; otherwise excessive green or stalky flavours will be absorbed.

11.7 Fermenting high-density musts to dryness

In hot climates, grapes are sometimes so sugar-rich that difficulties may arise in achieving fermentation to dryness. During the latter stages of fermentation, yeasts may suffer nitrogen starvation and ethanol stress. The problem may be overcome by the normally unmentionable and, in the EU, illegal addition of water to the must or in the early stages of fermentation. Winemakers who do this may refer to 'flushing out the picking bins very well' or using the help of 'the black snake' or 'the white horse'. These methods will, of course, reduce the level of alcohol in the finished wine. Without resorting to such tactics, techniques that may be employed include reducing fermentation temperatures, particularly in the later stages, and incremental additions of nitrogen.

11.8 Wine presses and pressing

We recall that in red winemaking, the press is normally used near, or after, the end of the fermentation to extract juice from the grape skins, and that in white winemaking the press is used before fermentation. For white wines, the grapes may initially be gently bruised or completely crushed before pressing. Alternatively, whole clusters of grapes may go straight to the press, which will result in a must lower in phenols.

There are many different types of press, and the type used will depend on a number of factors, as discussed below. A wine press may be categorised as either a continuous press or a batch press.

11.8.1 Continuous press

There are many variations of the system of continuous pressing, but all work on the principle that grapes are fed into one end, carried along the length of the press by an Archimedes screw, and are subjected to ever-increasing pressure as they continue their journey. Must is collected at various points, with better juice coming from the earlier collection points. However, none of the must will be of very high quality, and the use of such presses is not permitted for many Appellation Contrôlée (Appellation Protégée) wines. As long as grapes are being fed in, the machine will work. Continuous presses are mostly used in large 'industrial' wineries, having the advantage of efficiency in processing copious quantities of fruit.

11.8.2 Batch press

This category can be divided into three basic types: the horizontal plate press, the horizontal pneumatic press and the vertical basket press.

11.8.3 Horizontal plate press

This type is very often referred to as a 'Vaslin' press, being the name of the original manufacturer. NB: the present-day company, Bucher Vaslin, manufactures membrane presses and other winemaking equipment. A horizontal plate press consists of a slotted cylinder containing a system of plates, hoops and chains. Grapes are loaded into the press through a hatch, which is then closed, and the press rotates, tumbling the grapes. At this stage, high-quality free-run juice is extracted. With the press continuing to rotate, two metal plates move from each end along a screw towards the centre of the press, squeezing the grapes between them and releasing good-quality juice. The cylinder then rotates in the opposite direction, and the plates are returned to the ends of the press. For the next pressing, the press rotates, breaking up the 'cake' of pressed grapes, and the plates return to the centre, forcing out more juice, but of a lower quality. The press may operate three or more times in this way.

In each grape berry, the best-quality juice is in the area called *Zone 2* – this is neither close to the skin nor close to the pip, and it is this juice that is released in the first pressing. Plate presses are relatively inexpensive, and are particularly popular with smaller properties making red wines. An illustration of a horizontal plate press is shown in Figure 11.5.

11.8.4 Horizontal pneumatic press

This type of press is still sometimes referred to as a Willmes or Bucher press, again after the names of the original companies of manufacture. Pressing in a pneumatic press is a relatively gentle, low-pressure operation. An illustration of a pneumatic press is shown in Figure 11.6. The press rotates, and a sheet, bladder or central rubber bag is inflated by means of compressed air. The grapes are pressed against the entire slotted cylinder – a much larger surface area than in the plate press. Again, several operations will take place to extract all the usable juice, with the finest must being from the first pressing. A tank press is a fully enclosed pneumatic press, which can be flushed before use with an inert gas such as N_2 or CO_2, thus helping to eliminate the risk of oxidation of the must.

Pneumatic presses are particularly suitable when pressing grapes for white wine vinification, the gentle process helping to retain aromatics from the grapes. Nowadays, the press or even a whole bank of presses may be computer-controlled, adding to efficiency (and capital costs). Pneumatic presses are expensive, and they are only used for a few days in the year. Accordingly, the owners of small wineries are very keen to proudly display their recent purchases to any visiting wine lover who appears at all interested!

Figure 11.5 Horizontal plate press.

11.8.5 Vertical basket press

This is the traditional press, as developed by the monks of the Middle Ages. A metal (or still occasionally wooden) plate is lowered onto grapes contained in a basket, comprising either wooden staves or metal drilled with thousands of draining holes. A basket press is shown in Figure 11.7. There are several variations on the basic theme, including deep or shallow baskets, the latter still being extensively used in the Champagne region of northern France. Basket presses are very gentle.

The press can be used for whole clusters of grapes in white winemaking, or for the remaining solids for reds. The larger the diameter of the press, the more quickly the juice will run away from the bed of grape skins. The stalks help provide drainage channels. However, the draining juice is exposed to the

Figure 11.6 Pneumatic press.

air all along its route, and it is important to process it without delay in order to limit oxidation.

The use of basket presses is a particularly slow process, both in pressing and in cleaning. With horizontal presses, once the operation is complete, the cylinder is inverted, and the cake of dry skins is emptied through a hatch so the press can be cleaned. Vertical presses require the remaining skins and pips to be removed manually, and the basket must be cleaned thoroughly. Nevertheless, they are gaining in popularity again, particularly with many New World winemakers. One speciality use is for a particular style of red wine. Because of the nature of the pressing, the grape solids remain reasonably static, resulting in a press wine that may be less bitter than would be obtained from a pneumatic or plate press. This can allow the spicy flavours of certain varieties, such as Shiraz, to prevail. Small basket-type presses such as the Idro press, which works using a water-filled membrane, are particularly useful in small wineries, and for when micro-vinifications are undertaken.

Figure 11.7 Basket press.

11.9 Technology and the return to tradition

To visit many top-class wineries, or an equipment exhibition such as Vinitech-Sifel and Simei it is easy to be seduced by the latest highly technical winemaking equipment. Destemmers have had the same basic design for decades, but now there is Delta Oscillys, in which the berries are separated from the stems by inertia. Initially, sorting tables were a moving belt, later came vibrating tables, and more recently came the technology of optical sorting. Some properties employ all three of these! However, much technology in winemaking seems to be going full circle – as we have seen, many quality winemakers are now replacing their stainless steel vats with wood or concrete. There is also a revival in the usage of tinajas and qvevri (earthenware amphorae), the qvevri being buried in the ground.

CHAPTER 12

Barrel maturation and oak treatments

Many wines receive the influence of oak, during winemaking and/or maturation. In this chapter, we look at the history of the use of oak barrels, the influence of barrels upon wine, the impact of different types of barrel and the use of other oak products.

12.1 History of barrel usage

The earliest storage vessels for wine were amphorae, which began their use around 4800 BC, which is almost as far back as we can trace the beginning of winemaking. The history of wine shows that from early times, wood was used for the storage of wine. Various types of wood were experimented with, including palm wood used by the Mesopotamians. However, since Roman times, particularly since 200 AD, oak gradually became the most favoured choice. Winemakers from that time are known to have discovered that oak barrels, as well as a means of storage, made wine softer, and its taste was often improved. Wine began to be transported in barrels, which were much stronger than vessels made of clay. In the Rheinisches Landesmuseum in Trier, there is an exhibit of the tombstone of a wine merchant from Neumagen from c.220 AD, a wine ship containing four large barrels of wine.

Of course, barrels became the obvious type of container for exporting wine. As a result of the marriage in 1152 of Eleanor of Aquitaine to Henry of Anjou, who became king of England in 1154, most of western France including Bordeaux came under English rule. Bordeaux has an easy sea route to England, and a substantial export of wine developed. Tonneaux, barrels containing 900 litres of wine, were often used up until the sixteenth century. The load capacity

Wine Production and Quality, Second Edition. Keith Grainger and Hazel Tattersall.
© 2016 John Wiley & Sons, Ltd. Published 2016 by John Wiley & Sons, Ltd.

of ships became measured in tonnage, i.e. how many of these casks they could carry – 900 litres of wine weighs approximately 900 kg and the cask another 100 kg, equating to a metric tonne. However, tonneaux were very unwieldy and impractical for road transport, and by 1789 the use in Bordeaux of the 225-litre *barrique*, for both storage and transport, was the norm. The wine would be exported early, still on the lees, and often the barrels would not be returned, resulting in the use of a percentage of new wood for each new vintage.

12.2 Oak and oaking

Maturation in oak barrels has a major impact upon the style and quality of wines and is used for most classic, high-quality red wines. Many whites undergo fermentation and/or storage in barrels. The wood imparts oak products and flavours, including those from toasting of the wood, to the wine. The texture of the wine is changed, tannins may be softened, and the wood helps fix the colour. However, the back labels of countless inexpensive bottles contain such phrases as 'subtle hints of oak and vanilla' or 'oak-aged for complexity and richness'. Visions immediately come to mind of row upon row of barrels of wine gently maturing in cool, dark cellars, a truly impressive sight. However, inexpensive wines are very unlikely to be matured in oak barrels, for economic reasons. The cost of a new French oak 225 litre barrique, which holds 300 bottles of wine, is in the region of £450–£550. Accordingly, in the last 30 years or so, there has been substantial growth in the usage of oak products, such as chips and dust, to give wood influence to many wines.

12.3 The influence of the barrel

The influence of the barrel upon the style and quality of wine will depend on a number of factors including: size of barrel; type and origin of oak (or other wood); manufacturing techniques including toasting; amount of time spent in barrel; and where the barrels are stored. The age of the barrels has a major impact. When barrels are new, they impart many oak products to the wine, including vanillin, lignin and tannin. When they are used again (second fill), the amount of oak derived by the wine is reduced. After the barrels have been filled four or five times, they become merely storage containers that allow the wine to mature as a result of a very slow, controlled, oxygenation. This slow oxygenation softens tannins, causing them to precipitate and fall as sediment to the bottom of the barrel to be racked off later. However, oxygenation is greater in new barrels than those that have been filled several times, as with repeated use, the pores of the wood block become blocked with tartar and colouring matters. The absorption of oak products coupled with the oxygenation also helps fix wine colour.

The fine red wines of Bordeaux, Rioja and Burgundy (both red and white) are examples of wines whose character is developed by maturation in oak barrels. However, it is only high-quality wines that have the structure to be aged entirely in 100% new wood, and most producers will use a mix of barrels of different ages. Achieving a fine balance of oak aromas and obtaining the right level of oak influence are part of the art of a skilled winemaker.

12.3.1 Size of the barrel

The smaller the barrel, the greater the surface-to-volume (wine) ratio and thus the greater the influence of the oak on taste and also on the wine's maturation. A 225-litre barrique will give 15% more oak products to a wine than a 300-litre barrel. The smaller barrel will also help during fermentation in dissipating the heat produced. In some regions such as the Rhône Valley in France and Piemonte in Italy some wines are still matured in large oak barrels (botti/foudres) that hold 4000–6000 litres, imparting only minute amounts of oak. There are many sizes of barrels in use around the world, some having been used for centuries such as the 228-litre Pièce Bourguignonne and the 205-litre Pièce Champenoise.

12.3.2 Type and origin of oak (or other wood)

There are three main sources of oak that are commonly used in winemaking: America, France and Hungary. Although there are over 150 species of oak, only three of these are used to make wine barrels. American white oak – *Quercus alba* (Qa) is more coarse-grained than *Quercus robur* (Qr) (pedunculate or English oak), and the tightest-grained oak is *Quercus petraea* (Qp), also known as *Quercus sessiliflora* (sessile oak).

American oak is usually sawn, while with the tighter grains of French oak, the wood has to be split along the natural grain. Also, within France, there is a difference in the oak from the various forests: Allier, Limousin (Qr), Nevers (Qp), Jupilles, Tronçais (mostly Qp), Vosges, etc. It is not just the growing conditions that are different but also the weather conditions that affect the seasoning of the wood. Seasoning may take up to 4 years after the trees have been felled. Although descriptive bottle labels will sometimes indicate, for example, that the wine has been aged in 'Allier oak', winemakers now are more likely to choose individual coopers rather than particular forests. Most wineries use barrels from several coopers, giving a wide palette of components for the ultimate blends.

Forests cover some 25% of the land of France, some 14 million hectares, of which over 2 million hectares of Qp and Qr are in the regions used to source wood for wine barrels. The forests are generally well managed by the Office National des Fôrets (ONF), and the amount of oak wood in these forests has increased by 30% in the last 30 years. Hungarian oak is of the same species as much French oak – *Quercus petraea*. Although similar to French oak, the cost is much less. Other sources of oak include Russia, Slovenia and even China.

American oak imparts flavours fairly rapidly, including pronounced vanilla and coconut. Some grape varieties, e.g. Syrah (Shiraz) are particularly suited to American oak maturation, while others, e.g. Pinot Noir, may be overwhelmed with such loud flavours. French oak, especially Qp, gives a lower release of compounds. Woods other than oak have historically been used for barrels and are sometimes seen today. These include chestnut, acacia and, especially in Chile, rauli (*Lophozonia alpina*), a type of beech wood.

12.3.3 Manufacturing techniques including toasting

The techniques used in the coopering of barrels vary from country to country. Barrels made of American oak but coopered in France are very different from, and often preferred to, those coopered in the USA. Traditionally, in Europe, the wood is first split into staves and then allowed to rest outside for at least 2 years while 'seasoning' takes place, although three or more years' seasoning is specified by many discerning producers. An illustration of planks being seasoned is shown in Figure 12.1. The seasoning process allows harsh tannins to be drawn out. In the latter stages of manufacture, barrels will be 'toasted' to a greater or lesser degree, i.e. the insides, or maybe just the 'heads' (ends) will be charred over a small wooden fire. Winemakers will have a requirement for heavy, medium +, medium or light toasting according to the finished style required for their wine. Some of the flavours imparted by different levels of toast are summarised below:

- heavy toast: caramel, coffee, dark chocolate, toasted bread, smoky, spicy, cloves; less astringent tannins;
- medium toast: butterscotch, bread, vanilla, milk chocolate;
- light or soft toast: woody flavours, sometimes greenness, high wood tannins.

The type of wood used to fuel the fire will have an impact upon the toasting and consequently the wine stored in the barrel.

12.3.4 Stave thickness

Barrels are manufactured in several stave thicknesses, with 27 mm being the most common, as used for the so-called 'export' Bordeaux barriques. However, 22 mm staves as used for 'château' barriques give a slightly greater oxygenation, and some producers will use a percentage of 'château' barriques in their cellars.

12.3.5 Amount of time spent in barrel

Historically, wines were often kept in barrel until they were required for shipping – in some instances, this might have been as long as 10 years. Nowadays, it is exceptional for any wine to be kept for more than 3 years in wood. The longer the period, the more oak products will be absorbed, and the more oxygenation

Figure 12.1 Oak being seasoned for barrel-making.

takes place. The barrels will need topping up shortly after filling and in the early stages of storage, to prevent oxidation and bacterial growth. Initially, barrels will be stored with the bunghole on top, with the bung loosely inserted or with a fermentation lock in place. Later, the barrels may be stored with the sealed bung hole on the side, the contact with the wine ensuring that the bung remains expanded in the hole. During the first year in barrel, the wines may be racked four times, and during the second year, they may be racked twice.

During the time spent in barrel, there will be a small loss in the alcohol of the wine – depending upon the humidity and temperature of the cellar, the loss will be 0.2–0.8% abv over a 2-year period.

12.3.6 Where barrels are stored

Barrel maturation is an interplay between the wine, the wood and the atmosphere in which the barrels are stored. Wine evaporates through the wood and is replaced by oxygen, and oak products are absorbed. The process is temperature- and humidity-dependent. For example, in a warm, dry atmosphere, the volume of wine that evaporates is greater than in a cool, damp atmosphere. However, in a cool, damp atmosphere, a greater volume of alcohol evaporates. Thus, a wine from an individual property that is barrel-matured in an underground cellar may be subtly different from a previously identical wine that is matured in a warehouse above the ground.

12.4 Oak treatments

As we have seen, maturation of wines in new oak barrels is an expensive process not just because of the high cost of the barrels but also because of the extra labour in racking and working with small individual batches. Thus, inexpensive wines are likely to have been matured in vats, and if 'oaked', this usually involves oak products being soaked in the wine. Historically these simply comprised the waste from the barrel-making process, but now additionally many coopers undertake specialist manufacture of these oak products. The range includes oak dust, chips, cubes, spirals and staves. Of course, in the vat there is not the interplay between the wine, the wood and the atmosphere. Wines treated with chips often exhibit a sweet, vanilla sappiness. The level and style of oakiness depend upon the size and amount of the chips or cubes used (500 g/hl perhaps being an average), the source of the oak, how the chips have been toasted and, of course, the length of maceration time.

More subtle oaking can be achieved by the insertion of oak rods, planks or staves into vats of wine. The staves may be new and fire toasted, or may have been 'reconditioned' from broken down old barrels. They can be built into a tower or inserted into the tank as a module, rather like an inner lining to the sides of the tank. This system allows for a high level of extraction together with total control and, of course, with considerable cost savings over barrel maturation.

Even if bottle labels refer to 'barrel maturation', this does not necessarily mean that all the oakiness has come from the actual barrels. Oak chips or rods may be suspended in perforated sleeves through the bunghole and moved from barrel to barrel, as required. Alternatively, the barrels may have undergone a 'renewal system' – the insertion of new staves inside the barrel or the building of a grid of oak sticks.

Producing an inexpensive wine with subtle oak tones is certainly an art, and many different methods may be combined. For example, an unoaked vat component may be blended with a micro-oxygenated 'cubed' wine and perhaps a small percentage of wine that has been aged in second and third fill barrels.

CHAPTER 13

Preparing wine for bottling

In this chapter, we look at the various treatments that may be undertaken to ensure that wine is of the clarity expected and remains stable in the bottle for its projected life. These include fining, filtration and hot or cold stabilisation. Many makers of fine wines believe such treatments should be kept to a minimum, and there is constant debate among producers, critics and wine-loving consumers as to the organoleptic impact of the processes. The USA wine critic Robert Parker, who has been referred to as the *Emperor of Wine*, is well documented as preferring unfiltered wines.

Immediately after the conclusion of the alcoholic fermentation, wines will contain considerable solid matters including grape solids and yeast hulls. Given time, most of these solids will settle, and the wine may be racked off. We have already seen that racking is one of the operations necessary to achieve clear wine and to reduce the risk of off-flavours. It is important for the wine to be protected from oxidation during this racking process, and vats or barrels may be sparged with inert gases to achieve this.

13.1 Fining

Coarse matters that form sediments are removed by racking, or occasionally centrifuging. However, there may remain lighter matters suspended in the wine, known as colloids. These are capable of passing through any filter. If not removed, they may cause the wine to look hazy and eventually form a deposit. The colloids are electrostatically charged and can be removed by adding another colloid with the opposite charge. Molecules with the opposite electrical charge attract each other and form larger clusters that flocculate and

Wine Production and Quality, Second Edition. Keith Grainger and Hazel Tattersall.
© 2016 John Wiley & Sons, Ltd. Published 2016 by John Wiley & Sons, Ltd.

fall to the bottom of the vat or barrel, after which the wine may be racked of the sediment formed or can be filtered. The fining agent used will depend on the nature and effect of the colloids to be removed. Examples of fining agents for red wines include egg whites (albumin) and gelatine, both of which can remove some hard tannins and therefore reduce astringency. For whites, isinglass (obtained from swim bladders of fish) help remove astringency. Bentonite reduces protein, which is often responsible for hazy wines. Milk (casein) will lighten the colour and may help with minor oxidative taints – see Chapter 21. For phenolic compounds, gelatine may be used, and phenolics are also absorbed by the substance polyvinylpolypyrrolidone (PVPP). This product may also be used to remove colour from white wines and help prevent browning. Quantities of all fining agents need to be carefully controlled; otherwise the fining agent itself will form a deposit, or a further (opposite) electrostatic charge may be created.

13.2 Filtration

Filtration is the process used to remove solid particles and may take place at various stages in winemaking. Must may be filtered prior to fermentation; lees may be filtered to recover valuable wine; and filtration may take place prior to maturation in vats or barrels. However, most wines will be filtered during the preparation for bottling. It should be noted that the process of fining and filtering are not interchangeable: most filtration will not take out the colloids removed by fining, although it is claimed by some advocates, including equipment manufacturers, that membrane filtration eliminates the need to use of fining substances.

13.2.1 Traditional methods in common use

There are several methods of processing in common use in wineries around the world for the clarification and filtration of wine (and must and lees). Centrifugation, generally using a plate separator centrifuge, is effective for basic clarification. After fermentation, centrifugation may be used at various stages, including post-fining. It is often used pre-filtration; however, centrifugation is a violent method, and many wineries avoid the process. Models of centrifuge are now available that can be flushed with inert gas to avoid the problem of oxygen uptake. Alternatively, the wine may undergo depth filtration using kieselguhr (diatomaceous earth) powder. Diatomaceous earth has been used as a filtration adjuvant since the late nineteenth century. The filtration may be undertaken using a mobile earth filter or rotary vacuum filter as discussed below. Earth filtration is used for initial rough filtration and can remove large quantities of 'gummy' solids, which consist of dead yeast cells and other matter from the grapes. The filtration takes place in two stages. First, kieselguhr is deposited on a supporting screen within a filter tank. A mixture of water and kieselguhr may be used to develop the filter bed. This is known as pre-coating. Second, more earth is mixed with wine to form a slurry, which is used to continuously replenish the filtration surface through which the wine passes. Wine is passed through the filter, and the bed gradually increases in depth. Eventually, it will bind, and the kieselguhr will have to be completely replaced with fresh material.

Another method is the use of a rotary vacuum filter, which consists of a large horizontal cylinder or drum with a perforated screen covering the curved surface. A filter cloth is stretched over the surface. The cylinder rotates in a trough containing kieselguhr and water. A vacuum is drawn on the cylinder, and the kieselguhr–water mix is drawn on to the cloth. The water is drawn into the cylinder leaving a layer of kieselguhr on the cloth to act as a fine filter medium. Wine is fed to the trough and filtered through the kieselguhr. As the surface fouls, more kieselguhr can be added to the wine to replace the fouled layer, which is scraped off by blades as the cylinder rotates. Figure 13.1 illustrates, a pair of rotary vacuum filters. A sophisticated filter is

Figure 13.1 Two rotary vacuum filters.

the totally enclosed rotary vacuum filter, which reduces the risk of oxidation damaging the wine. However, with this type of filter, the fouled earth has to be removed manually.

Diatomaceous earth is a carcinogenic material and presents issues regarding deposits of residue in landfill sites. Perlite, made from processed volcanic rock, may also be used for filtration of grape must and wine containing a high percentage of solids.

13.2.2 Sheet filtration (sometimes called plate filtration)

Although they will not cope with wines containing gross solids, a sheet filter, as illustrated in Figure 13.2, is very versatile and extensively used in wineries of all sizes. A sheet filter may also be referred to as a pad or frame filter. In small wineries, it may be the only type of filter available. A series of specially designed perforated steel or plastic plates are held in a frame. Sheets of filter medium are placed between the plates, which are then squeezed together by an Archimedes screw or by hydraulic methods. Filter sheets are available with various ranges of porosity. Wine is pumped between pairs of plates to pass through the filter sheets into a cavity in the plates and then to exit the system. Yeast cells and other matter are trapped in the fibres of the filter media. The sheets are constructed of cellulose fibres, sometimes with the addition of granular

Figure 13.2 Sheet filter.

components, such as kieselguhr, perlite or polyethylene fibres, and sometimes also cation resins (which have an electrostatic charge) to attract particles with the opposite electrostatic charge. Filter sheets are manufactured in several grades with uses ranging from basic polishing to complete sterilisation. However, the filter sheets gradually become blocked, necessitating a labour-intensive and time-consuming dismantling and reassembly of the machine. This operation also results in wine loss. The filters can also be messy – it is a very rare exception to see a drip-free machine. The enclosed lenticular depth filter is a more hygienic and less labour-intensive piece of equipment.

13.2.3 Membrane filtration and other methods of achieving biological stability

Membrane filtration of wine can be carried out as front-end membrane filtration, which necessitates previous clarification and filtration using the methods detailed above, or cross-flow filtration, which will achieve the desired clarity in a single pass. A cross-flow filter is illustrated in Figure 13.3.

Figure 13.3 Cross-flow filter.

If a wine contains any residual sugar and has an alcoholic degree of less than 15.5% abv, there is a considerable risk of a most undesirable re-fermentation taking place in bottle (or other packaging), if any yeast colony-forming units are present. Thermotic bottling, heating the pre-bottled wine to 54°C (129°F), is one method commonly used to guard against this risk, but it does present issues with fill heights in bottles as the wine cools. Other methods are flash pasteurisation, immediately prior to bottling, at a temperature of approximately 75°C (167°F) for 30 s and post-bottling tunnel pasteurisation at 82°C (179.6°F) for approximately 15 min. However, membrane filtration immediately prior to bottling is the method preferred by many quality-conscious producers. It avoids flavour modifications occasionally associated with pasteurisation. Membrane filtration is also a sensible precaution prior to bottling of wines that have not undergone MLF, as the consequences of this taking place in bottle would also result in wine spoilage. *Leuconostoc oenos*, *Pediococcus damnosus*, *Oenococcus oeni* and *Lactobacillus brevis* (lactic acid bacteria) are the main bacteria involved and are all removed by a 0.45 μm membrane. *Lb. brevis* can break down compounds in wine and result in, inter alia, a threefold increase in volatile acidity. Yeasts of the species *Saccharomyces cerevisiae* (the most dominant yeast in most fermentations) and the closely related *Saccharomyces bayanus* are removed using a 0.8 μm membrane pore size. *Dekkera intermedia,* a sporulating form of *Brettanomyces*, which causes a defect recognisable on the nose as a pronounced 'burnt sugar' smell or a 'mousey' off-flavour are removed at 1 μm. Unenclosed plate filters, of course, may permit immediate reintroduction of yeasts and bacteria. Membranes with pore diameters of 0.2–0.45 μm are usually used for filtering white wine and 0.45–0.65μm for reds.

13.3 Stabilisation

Stabilisation may be carried out to reduce the likelihood of tartrate crystals forming in the wine after bottling. The tartrates are either potassium or calcium salts of tartaric acid and are totally harmless. If present, they may attach themselves to the cork or fall as sediment in the bottle, particularly in white wines which, of course, have higher levels of tartaric acid than reds and are stored cold in restaurants and bars. The presence of such tartrates sometimes cause unwarranted concern to consumers.

Some tartrates will be removed during the racking operations post-fermentation. Figure 13.4 shows tartrates left in a tank after a red wine has been racked. The conventional methods of inhibiting the precipitation of tartrates in the finished wine are by cold stabilisation or contact process shortly before bottling. Using cold stabilisation, the wine is first fined to negate the protective effect of micro-crystals. It is then chilled to just above its freezing

Figure 13.4 Tartrates left in a tank after racking.

point, −4°C (24.8°F) for a wine containing 12% abv and −8°C (17.6°F) in the case of fortified wines containing 16–22% abv. The wine is held, preferably in insulated tanks, at this temperature for at least 8 days and possibly considerably more. After this operation, the wine can be racked off the crystals that will have formed, and can be bottled.

However, cold stabilisation for this purpose is expensive in terms of both equipment and running cost of the refrigeration units, and not always effective in the long term. A more effective, quicker and cheaper system is that of the contact process. This involves chilling wine to 0°C (32°F) and adding 4 g of finely ground crystals of potassium bitartrate ($KC_4H_5O_6$) per litre of wine, followed by a vigorous stirring to keep them in suspension. Alternatively, the wine is pumped through the crystal bed. This method will take 5 days or so. Ion exchange, replacing potassium and calcium ions in wine with sodium ions, is perhaps, from a health point of view, undesirable. The process was until recently banned in the EU, but is now permitted under Regulation (EC) No. 606/2009. The addition of metatartaric acid or carboxymethylcellulose is permitted in the EU up to 100 mg/l. However, such addition is only effective at protecting from tartrate precipitation for a limited amount of time, perhaps 9 months or so. A further alternative is the addition of mannoproteins, which are extracted from the cell walls of yeasts, but they react with cellulose, and therefore if used, the addition must be prior to sheet filtration.

A far more effective method of tartrate stabilisation is by the use of electrodialysis, but the capital cost of the equipment puts this out of reach of many wineries, although, in some regions, mobile machines and operators can be hired. Electrodialysis uses selective membranes allowing the passage of potassium, calcium and tartrate ions under the influence of an electrical charge. The removal of tartrates by this method is fast (a single cross-flow pass), efficient and reliable, and is adjusted to the characteristics of the wine being processed. The speed of the process makes the system particularly attractive to producers who wish to get a wine, such as a new season's Sauvignon Blanc, to market early. There is also less wine loss, perhaps 1%, compared with up to 3.5% with cold stabilisation. As the process takes place without refrigeration, there is a considerable reduction in energy costs, up to 95% compared with the 8+ days for cold stabilisation.

13.4 Adjustment of sulfur dioxide levels

As we have seen, sulfur dioxide (SO_2) is the winemaker's generally used antioxidant and disinfectant. Before bottling, the free SO_2 levels should be adjusted, usually to between 25 and 40 mg/l. However, careful winemakers will relate the desired level to, inter alia, the pH of the wine, with lower pHs requiring less. For example, a red wine with a pH of 3.1 might only require 16 mg/l for protection; at a pH of 3.9, the figure would be 99 mg/l, well above the sensory threshold. Sweet wines need higher levels to inhibit further fermentation of the residual sugars (RS), although potassium sorbate ($CH_3CH=CH-CH=CHCOOK$) is most effective as a fermentation inhibitor. The maximum permitted level for potassium sorbate under EU regulations and Australian regulations is 200 mg/l. In the USA [Bureau of Alcohol, Tobacco, Firearms and Explosives (BATF)], it is 300 mg/l. Potassium sorbate is not a permitted additive for producers seeking organic certification. The maximum permitted levels of total SO_2 for wines produced or marketed in the EU are shown in Table 13.1. In Australia, the figure is a total of 200 mg/l for a wine with less than 35 mg/l RS, and 300 mg/l if there is more than 35 g/l RS (Standard 4.5.1). In the USA (BATF), the total permitted SO_2 is 300 mg/l irrespective of the level of RS.

13.5 Choice of bottle closures

Corks are the traditional way of closing wine bottles and still account for approximately 12 billion of the 20 billion closures produced per annum. The largest producer is Amorim, which, in 2014, had sales of over 4 billion cork

Table 13.1 EU permitted levels of SO_2 in wine

Dry red wine	150 mg/l
Dry white or rosé wine	200 mg/l
Red wine with 5 g/l sugar or more	200 mg/l
White wine with 5 g/l sugar or more	250 mg/l
Certain specified white wines, e.g. Spätlese	300 mg/l
Certain specified white wines, e.g. Auslese	350 mg/l
Certain specified white wines, e.g. Barsac	400 mg/l
'Quality' sparkling wines	185 mg/l
'Other' sparkling wines	225 mg/l

Commission Regulation (EC) No. 606/2009 Annex 1 B.
NB: lower limits apply to wines labelled 'organic'.

closures. However, other types of closure have gained (and in some cases waned) in popularity in the last 20 years. Cork is a natural product from the bark of a type of evergreen oak tree, *Quercus suber*. Portugal produces over half the world's cork.

A very important quality of cork is its elasticity. When compressed, a cork will try to regain to its original size. Hence, when used as a closure for wine bottles, it will provide a tight seal so long as it is kept moist. Wine bottles sealed with a cork closure should therefore be kept laying on their sides to prevent the cork from drying out. However, because corks are a natural product, the quality is variable and sometimes a taint can be imparted to the wine, as discussed in Chapter 21.

Various synthetic materials have been developed as an alternative to natural cork. Plastic stoppers have now been on the market for some years but are still proving controversial. A common consumer complaint is of the damage that plastic stoppers can do to corkscrews. Also, the plastic will harden over time, allowing oxygen ingress into the wine owing to the seal against the bottle neck failing. Aesthetically, there can be no doubt that these are not natural corks and indeed some producers choose vivid colours for the stoppers to complement the labelling. Diam is a moulded closure made from particles of natural cork that have been flushed with supercritical carbon dioxide (which, it is claimed, eliminates taints above the sensory threshold), together with microspheres. The metal screw cap is regarded by many winemakers as the best means of ensuring wines that are haloanisole taint-free and preserving maximum fruit and freshness. The Amcor brand Stelvin® is the world market leader. It should be noted that it is not the screw cap itself that seals the wine, but the liner, which is made from Saran™, which is polyvinylidene chloride. There are basically two types of liner: Saranex, which comprises multiples layers of Saran™, and Saran/Tin, which, as the name suggests is a Saran™ laminate on a thin layer of tin. Saran/Tin has the greatest barrier to oxygen permeation. Normacorc® is a closure with a foam core with elastic outer skin. At the time of writing, some 2.4 billion bottles a year are sealed with this closure. Vinolok is basically a closure made of glass and marketed as such. However, the actual seal between the stopper and bottle is made from an ethylene and vinyl acetate copolymer.

The impact of various types of closure upon the bottled wine is the subject of ongoing research. One topic of particular concern is that of 'flavour scalping', i.e. the absorption of aromas and flavours by the closure.

CHAPTER 14

Making other types of still wine

So far, we have discussed the procedures for making red and dry white still wines. In this chapter, we will now look very briefly at ways in which other styles of wines are made. We will briefly consider the making of sweet wines, wines made from semi-dried grapes, rosé and fortified wines. We will touch on some individual wines produced by these methods in various parts of the world.

14.1 Medium-sweet and sweet wines

Generally speaking, following natural fermentation, most or nearly all of the sugars will have been converted by the yeast. Any sweetness of the wine is determined by the amount of residual sugar. The EU has legal definitions for the description of the terms sweetness and dryness. For example, wines containing less than 4 g of residual sugar per litre are described as 'dry'. At the other end of the scale, a wine described as sweet must have no less than 45 g of residual sugar per litre.

If medium-sweet or sweet wines are required, there are several ways they can be produced. The method used will depend upon the ripeness of the fruit available, and the style, quality and price of the wine aimed for. Fermentation can be stopped deliberately while a quantity of sugar remains in the wine. Historically winemakers would use a heavy dose of SO_2 to kill off the yeasts. Nowadays, the quantity used is strictly controlled by law. One method of stopping fermentation is by cooling the wine so that the yeast becomes dormant and inactive. The wine will then undergo filtration to remove the yeast cells. Another way is for the alcohol level to be raised by the addition of grape

Wine Production and Quality, Second Edition. Keith Grainger and Hazel Tattersall.
© 2016 John Wiley & Sons, Ltd. Published 2016 by John Wiley & Sons, Ltd.

spirit (as in Port winemaking). Depending on the yeast strain, 15% alcohol is about the maximum level to which fermentation is possible.

Particular techniques can be used for making medium-sweet or sweet wines as discussed briefly below.

14.1.1 Medium-sweet wines

These can be made by the addition of a small amount (10–15%) of sterile unfermented grape juice once the wine has finished fermentation. This is a practice commonly followed in Germany for inexpensive wines, where the juice added is known as süssreserve. This helps reduce any aggressive acidity (and the alcoholic degree) and helps maintain fresh grapey flavours.

14.1.2 Sweet wines

In certain vineyards and/or in exceptional vintages grapes can ripen to such an extent as to have very high sugar levels. Provided the weather is favourable, the grapes may be left on the vine until well after normal harvest time with the aim of increasing both sugars and fruit concentration. Such wines may be marketed as 'late harvest' or, in Germany and Austria, as 'spätlese'. Alternatively or additionally, the ripest bunches may be selected and an 'auslese' (bunch selection) wine produced. Such wines are often, but not always, sweet.

14.1.2.1 *Botrytis cinerea* as 'noble rot'

In Chapter 5, we noted that botrytis can appear in a welcome form as 'noble rot' when certain climatic and weather conditions occur. The ideal conditions are damp, misty early autumn mornings giving way to very warm, sunny afternoons. Vineyards close to rivers, such as the Layon in the Loire Valley in France and the Mosel in Germany may have suitable mesoclimates. A particularly gifted location in Bordeaux is at Sauternes, where the cold waters from the little River Ciron flow into the warmer waters of the Garonne, thus creating mists.

The grape varieties Sémillon, Riesling and Chenin Blanc are particularly susceptible to noble rot. The fungus gets to work by attacking the inside of the ripe grapes, which may contain approximately 200 g of sugar per litre. The fungus consumes sugars and effects chemical changes. It then attacks the skins, which become thin, fragile and permeable, and water content evaporates, thus concentrating the sugars and juice. The skins take on a brown/plum colour, and the grapes, which now contain approximately 250 g of sugar per litre are known as 'fully rotted'. In a further stage, the grapes will dry more and appear wrinkled. This is perhaps the state of perfection, and is known as confit or roasted rotted (pourri rôti). Most grapes are picked at this stage. The sugars equate to approximately 300 g/l. These sugars levels cannot be achieved in normal ripening. With further attack the grapes may develop a state of 'old roasted rotted' (pourri rôti vieux), appearing like raisins, with

440 g of sugar per litre or more, and the inclusion of these grapes will add a particular richness. The individual berries within a bunch may well be at different stages of attack. Thus, the pickers (usually local people who are experienced at selecting the desired berries) have to return to the vineyards several times in a series of successive pickings in order to select the roasted, rotted grapes. There may be five to nine selected passes, and this can take up to 2 months. This process is, of course, extremely costly. The grapes must be picked when dry, and the harvest sometimes continues until late November in the northern hemisphere. The quality and quantity of vintages vary considerably – the crop is at the mercy of the weather, and in some years *Botrytis cinerea* does not develop successfully.

Just as yeast struggles to work at low temperatures, high sugar levels can reduce the yeast's alcohol tolerance. Thus, fermentation will stop before all the sugar has been converted. Both high alcohol levels and high residual sugars can be achieved, as in in Sauternes, for example, which contains around 13–14% abv and sugar levels well in excess of 150 g/l. However, some very sweet German wines, such as Beerenauslesen and Trockenbeerenauslesen, may ferment to only around 8% abv. The cold winter will lower the cellar temperature so the yeast will cease to work, consequently leaving very high sugar levels.

Let us consider the making of a typical Sauternes: Château Bastor-Lamontagne. The yield is low, usually 18–20 hl/ha, but in 2014 it was a mere 13 hl/ha. The grapes are selectively picked in series of pickings and pressed in a horizontal press. The first pressing gives must with the most aromatics and the third (last pressing) juice with the highest sugars. The pressings are blended according to the winemaker's assessment. Natural yeasts are used, and even though the fermentation temperature is 20–22°C, this will take 3–6 weeks, depending on the richness of the must. The relatively high fermentation temperatures help give the wine body and richness. All the fermentations are started in stainless steel vats. The fermenting wines are then transferred to barriques to continue, with 15% of these being new. During the fermentation, it may be necessary to heat the barrel cellar. When the winemaker considers that the wines have the correct balance of alcohol and sugar, the fermentation is stopped by chilling the barrel room, transferring the wine into vats to be further chilled to 5–6°C (41–42.8°F) and adding SO_2. Maturation takes place in barrel for 14–16 months, with the components being blended shortly before bottling.

14.1.2.2 Appassimento

Traditionally in some parts of the world, grapes are dried or 'raisined' as a means of concentrating sugar levels. This may take place on grass mats in the fields under the sun, or alternative practices are carried out in some areas. For example, in Veneto in Italy, Amarone and Recioto Della Valpolicella are made from grapes that have been laid on trays or hung from rafters in ventilated or

Figure 14.1 Grapes undergoing the appassimento process.

air-conditioned rooms. During the 4 months or so that the grapes are stored this way, they lose up to 50% of their water content, thus resulting in a concentration of sugars. It is important that such grapes are healthy and free of botrytis. With such high sugar levels, fermentation is a very slow process and can take months. Grapes for Amarone undergoing the Appassimento process are shown in Figure 14.1.

Another means of concentrating sugar levels is to allow the grapes to remain on the vine, in anticipation of late autumn or early winter freezing conditions. Eisweins from Germany and Austria are made in this way. These grapes are not usually affected by *Botrytis cinerea* and will be harvested at temperatures below –7°C (19.4°F). Following pressing, the frozen water content of the grapes is discarded, and the syrupy concentrated juice is drained for fermentation. Canada also produces an Icewine.

- Tokaji Aszú – This is one of the world's best-known sweet wines from northern Hungary, made from Aszú (botrytis-infected grapes). In late October, botrytised grapes along with unaffected grapes are harvested at the same time but kept separate. The healthy grapes are made into a dry white wine, but the botrytised grapes are stored. Their high sugar content makes fermentation difficult. These grapes are then pounded into an Aszú paste and added to the dry wine to achieve the desired sweetness.

- Vin Santo – lit. holy wine – This is an Italian sweet wine, traditional in Toscana but also made elsewhere in Italy. It is usually made from white grapes, although a rosé style is sometimes produced using red grapes such as Sangiovese. When harvested, grapes are dried on straw mats in well-ventilated rooms or by hanging on racks inside.

14.2 Rosé wines

The popularity and sales of rosé wines have risen substantially over the last 10 years. Once seen as a summertime drink, the wide range of styles now available to the consumer from producers around the world has shown that this is no longer so. Rosé wines are appearing on the wine lists of highly rated restaurants, as the wines have good food and wine pairing qualities.

Rosé wines can vary in colour from the palest of pinks to light red, depending on the grape varieties, climate of the region and winemaking methods. The sweetness levels of rosés can range from dry, off-dry to medium-sweet. The wines can be made in still, semi-sparkling or sparkling styles. Alcohol levels vary considerably, too, depending on grape ripeness and winemaking techniques. In France, for example, the wide diversity of rosé styles range from Champagne and the Loire wines in the cool north, down to Bordeaux, Beaujolais, Rhône, and Provence and Languedoc-Roussillon in the south.

'Blush' is a term for the palest rosés that became popular in California in the 1980s. The Spanish and Portuguese use the term 'rosado', and the Italians use the terms 'rosato' and 'chiaretto'. 'Clairet' is a style of deep rosé from Bordeaux.

There are several methods used for producing rosé wines, but almost all will be fermented in tank and bottled young.

14.2.1 Blending

The blending of red and white wine together is now permitted in the EU. Historically, this method could not be used, except in the Champagne region where it has long been used for the production of rosé Champagne.

14.2.2 Skin contact

This style of rosé is made from red grapes or white grapes with heavily pigmented skins, such as Pinot Gris. Some grapes are crushed, skin contact may take place for 4–16 h, and then the grapes are drained. Other grapes are pressed. The drained juice together with the lightly pressed juice is then fermented as white wine without the skins. Pigments from the skins give some degree of colour to the juice. Pale rosés are often made by this method, and these may be labelled 'vin gris' (grey wine), although they are not in fact grey. Many rosé wines from the Loire are made in this way, as are California's 'blush' wines. If extra colour is required, a little red wine may be blended in.

14.2.3 Saignée

This French term means 'bled'. The wine may be made by a short fermentation on the grape skins. The maceration may take 6–16 h; then, the juice is drained, and this will be fermented cool. It is also possible to produce a rosé 'by-product', by draining part of a vat for rosé wine and leaving behind a greater concentration of skins to produce a full-bodied red wine. Alternatively, red grapes can be chilled and lightly crushed just sufficiently to release their juice. They are then allowed to macerate for 6–24 h, during which time the colour pigments and flavourings will 'bleed' into the juice. The juice is then drained off, and fermentation begins. This method can produce a deeper rosé colour, especially if the grapes have dark skins, e.g. Cabernet Sauvignon. This style of rosé is likely to be firmer, often showing some tannins.

14.3 Fortified (liqueur) wines

This category of wines has higher alcohol levels than the light wines previously discussed. Indeed, they are called 'fortified', since a quantity of grape spirit is added to 'fortify' them. The resulting alcohol levels in these wines can vary between 15% and 22% abv. Historically, brandy was added to light wine in barrels as a means of preservation and stabilisation, especially on long sea voyages. Detailing the production of fortified wines is an extensive subject, but we will consider the topic very briefly here. There are two basic methods of production of fortified wines. One method of production of liqueur wines involves fermentation to dryness before the addition of grape spirit. The other basic method involves the addition of grape spirit during the fermentation process, taking the alcoholic degree above that at which yeasts will work, resulting in naturally sweet wines.

The most well-known examples of fortified wines are Sherry and Port. A crucial difference in the method of making the wines is the timing of the addition of the grape spirit. Sherry is initially fermented to dryness; Port has the spirit added during the fermentation.

14.3.1 Sherry production

Sherry is the fortified wine produced in the delimited zone of Jerez (Andalucia) in south-west Spain. This is a relatively small region that lies on undulating chalk hills facing the Atlantic, and so the very hot summers are cooled by the sea breezes. The wine is believed to have been produced here for about 3000 years. Of the three soil types in this region, the most important is the albariza, which has the highest chalk content of 60–80% and is consequently very absorbent. The three main grape varieties are Palomino Fino, Pedro Ximenez (PX) and Moscatel. However, Palomino is by far the most important variety and the single variety responsible for the base wine.

PX is used for sweetening as well as some rare varietals of its own, and Moscatel is used to make its own varietal.

14.3.1.1 The ageing process

Once fermentation has ceased leaving the wines fully dry, they are classified as either 'Finos' or 'Olorosos'. The wines will have an alcohol level of 10.5–11.5% abv. Wines are then transferred into American oak barrels known as 'butts'. The butts are filled to about five-sixths, leaving an air gap over the top of the wine. A feature of the Sherry-producing region is a naturally occurring yeast (*Saccharomyces beticus*), which may grow on the surface of wines in barrels or vats, creating a yeasty crust known as 'flor'. Flor tends to grow thickest in cooler coastal areas, such as Sanlucar de Barrameda. Here, Manzanilla Sherry in cask has active flor growth on the surface of the wine for 12 months of the year. The bodegas in this region are usually large and very high. The height is to encourage cooling sea breezes, through the openings situated high up on the walls thereby keeping the temperature down. The colour and characteristics of Sherry will be dependent on the level and length of flor growth. Manzanillas and Finos from Sanlucar and Puerto will be a pale lemon colour, with very little oxidation, whereas Finos from Jerez, where the flor dies back in summer and winter, will be more golden.

After fermentation, wines destined for the Fino style are lightly fortified to 15%, which is the ideal alcoholic degree at which flor develops. Flor also requires oxygen and nutrients to live – these are naturally present in the wine. The crust of flor protects the wines from oxidation – the ageing of the wines is microbiological. If flor dies back, microbiological ageing will cease, and chemical changes will take place. Oxidation will occur, and the wines will naturally darken. These wines may then be designated as 'Amontillados'.

Wines of the heavier Oloroso styles will have been fortified up to 17% or 18% after fermentation. Flor will not grow at these alcoholic degrees, and so the wines will age in an oxidative manner. They may ultimately be sweetened as for 'cream Sherries'. The most important factor in the production of Sherry wines is the method of ageing used, known as the 'solera system'. This is a means of blending young wines with old to ensure consistency of style and quality. Each style of wine has its individual 'solera'. The solera comprises a stock of oak casks, known as 'butts', which are stacked in rows on top of each other, according to their stage of maturation, and called 'criaderas' (lit. nursery). The oldest row is actually named the 'solera'. The wines for bottling are withdrawn from the 'solera', which is then topped with younger wines from the 'criaderas', in order to replace and refresh. This is known as 'running the scales'. Prior to bottling, filtration will, of course, take place.

In order to be legally sold as Sherry, the wines have to be aged for at least 3 years in the solera system. Recent developments have brought a new, unusual style of Sherry to the market called 'en rama', literally meaning 'raw'. The wine does in fact receive a minimal filtration but retains a fresh style. Most bodegas now produce a limited quantity of this style of Sherry.

14.3.1.2 Age-dated Sherries

Although Sherry is a non-vintage wine, in July 2000 the regional Consejo Regulador, which is the regulatory public body representing all interests within Sherry production, created two special categories to award certain very old wines accreditation. These give an indication of their approximate age and quality:

- V.O.S. (Vinum Optimum Signatum or Very Old Sherry) – over 20 years old;
- V.O.R.S. (Vinum Optimum Rare Signatum or Very Old Rare Sherry) – over 30 years old.

More recently, the age certification system is available to those Sherries which although not attaining the age of V.O.S. or V.O.R.S. category, have had long periods of ageing and are of the highest levels of quality. Wines of 12 and 15 years have now been added to this category.

14.3.2 Port production

The Upper Douro River in Northern Portugal is home to the vineyards producing grapes for Port wine, one of the oldest demarcated wine regions in the world. Vineyards are on both sides of the river. This inland region, well away from Atlantic sea breezes, has very hot, dry summers with cold winters. The shallow soil is mainly schist with some granite.

While the list of permitted grape varieties is lengthy, just five varieties are principally used: Touriga Nacional, Tinta Roriz, Tinta Barroca, Tinta Cão and Touriga Franca (formerly known as Tinta Francesa). A most important feature of Port making is the extraction of colour and tannin from the grape skins. Historically, this was done by treading the grapes by foot, perhaps the gentlest way. More mechanised methods are now used. During the early stages of fermentation, it is critical that the skins be kept in contact with the juice to continue the extraction of colour and tannins. The fermentation process is then interrupted at an alcohol level of 6–9% abv, according to the house style. The juice is drained off and fortified to 18–20% abv. Yeasts will not work at this level of alcohol, and so fermentation stops. Sugar levels of about 100 g/l will be retained. There are various styles of Port including ruby, tawny, Late Bottled Vintage (LBV) and Vintage. Each of these styles has its own particular length and style of maturation. Some styles involve decades of barrel maturation; others are bottled early and need decades in bottle. Some Port houses also produce a 'single Quinta' wine, bottling a wine from their best estate(s) when a vintage is not being declared that year. Some white Port is also made using various white grape varieties such as Malvasia Fina and Codega. Styles of white Port range from dry to sweet.

14.3.3 Other well-known fortified wines

Fortified wines are made in many wine-producing countries, using an array of production methods. The following are just a few well-known examples.

14.3.3.1 Madeira

This is produced in a range of styles from dry to sweet. Drier styles are fortified after fermentation, while sweeter styles are vinified in a manner basically similar to Port. Traditional white grape varieties used are Sercial, Verdelho, Boal and Malvasia. In fact, most Madeira is made from red grapes – Tinta Negra is the principle variety. After fortification and clarification, the wines undergo a protracted heat treatment, either in estufas (heated tanks) or by the *canteiro* process, which is used for all the white wines. This involves storing barrels of wine for two years or more. During the heating of the wines, they gain oxidative characteristics, and the complex chemical changes (together with the wines' natural high acidity) result in Madeira wines having tremendous longevity.

14.3.3.2 Marsala

Marsala, from Sicily, is a fully fermented wine that is fortified with grape spirit and usually sweetened with boiled down must (known as mosto cotto). Wines are classified according to their colour and ageing. Older wines may be aged in a solera system. The oldest wine, 'Vergine', is aged for a minimum of 5 years and is unsweetened.

14.3.3.3 Malaga

This classic wine from Spain is vinified in a similar manner to Port but is often aged in the solera system as for Sherry.

14.3.3.4 Vins doux naturels

This is a range of 'naturally sweet wines' produced in many parts of southern France. These are mainly white, made from varieties of Muscat grapes, such as Muscat de Beaumes-de-Venise in the southern Rhône. There are, however, a few reds from south-west France made from Grenache Noir, for example, Banyuls in Roussillon. They are naturally sweet, having had the fermentations arrested by the addition of alcohol as for Port.

14.3.3.5 Commandaria

Commandaria is produced in Cyprus. After picking, the grapes are dried on mats in the sun until they have been 'raisined'. Fermentation usually stops before all the sugar has been consumed, leaving a naturally sweet wine. Following this, grape spirit is added to give a final alcohol of 20% abv.

14.3.3.6 Moscatel de Sétubal

Moscatel de Sétubal is produced in the Sétubal Peninsular in Portugal in a similar manner to Port. The grapes used are mainly Muscat of Alexandria. After pressing, the must is fermented for a short time and then stopped by the addition of spirit. The wine is then transferred to large vats, where fresh Muscat grape skins are added, and then left to mature over winter. A more pronounced Muscat character results.

14.3.3.7 Vin Jaune

Vin Jaune is produced in the Jura region of eastern France. Vin Jaune is made from the Savagnin grape, late harvested to give potentially 13–15% abv. Following slow fermentation, the wine matures in small oak barrels on ullage for at least 6 years, with no topping up. Consequently, a film of yeast grows over the wine, giving some protection from oxidation. This is similar to the flor in Sherry and known in the Jura as volie (veil).

CHAPTER 15

Sparkling wines

The bubbles in all fizzy drinks, including sparkling wines, are carbon dioxide (CO_2). In the case of most drinks other than wine (and bottle-conditioned beers), CO_2 is injected into the liquid before bottling. While it is true that some wines (e.g. most inexpensive Lambruscos) are given a spritz by this injection method, the sparkle in all quality sparkling wines is the result of retaining the natural CO_2 produced by a fermentation in the wine. The way that this fermentation is carried out, impacts upon the style and quality of the finished wine. Styles can vary from the lightest, foamy, sweet Lambrusco to the creamy, elegant mousse and biscuity tones of a bone-dry Champagne. In this chapter, we consider the different basic methods for producing sparkling wines and look in more depth at the 'traditional method', used for Champagne and other high-quality wines.

There are two basic methods by which this is achieved: *fermentation in sealed tank* or *fermentation in bottle*. The method of fermentation is the key factor in determining the finished style of sparkling wine.

15.1 Fermentation in a sealed tank

In order to retain carbon dioxide in the wine created by fermentation in a tank, this must be sealed and pressure-resistant. There are two variations in the process, according to the style of wine required. The first is the *continuous method*, a single fermentation as in the case of Asti and Prosecco in northern Italy. The latter, once produced in a similar style to Asti, has in recent years undergone a considerable change in styles and considerably increased

---

Wine Production and Quality, Second Edition. Keith Grainger and Hazel Tattersall.
© 2016 John Wiley & Sons, Ltd. Published 2016 by John Wiley & Sons, Ltd.

production. Prosecco is now produced as both fully sparkling (spumante) and lightly sparkling (frizzante).

The second tank method is the *charmat method*, also known as *cuve close*, which involves a second fermentation in a stainless steel sealed tank under pressure. This method is used widely in the production of New World sparkling wines. The dry base wine with the addition of yeast and sugar is re-fermented in the sealed tank, following which the wine is cooled. After this second fermentation, the yeast cells die. While the wine is in contact with the yeast, a complex chemical reaction (yeast autolysis) takes place, whereby the enzymes interact with the wine. The effect of this can add to the wine's complexity, aromas and flavours, giving it a creamy mouth feel and fuller body. Although not appropriate for all wines, it is recognised for adding a distinctive character to quality sparkling wines. The dead yeasts are removed by filtration, and the wine is bottled under pressure.

15.2 Second fermentation in bottle

Wines may be made sparkling by a second fermentation in a bottle. There are two variations: the *traditional method*, a second fermentation in the bottle in which the wine is sold, as in the case of Champagne; and the *transfer method*, a second fermentation in a *cellar* bottle (not the bottle in which the wine is sold), as in the case of many Australian sparkling wines.

As we have already seen, fermentation of wines depends on yeast cells in suspension. When the fermentation is finished, the yeast will fall as lees (dead yeast cells) to the bottom of the vat or other vessel, unless agitation takes place to keep it in suspension. Before the making of a sparkling wine is finished, the yeast or yeast sediment must be removed. In the *charmat* method, this is simply a matter of filtering out the yeast and bottling the wine under pressure. In the *transfer method*, the bottles are decanted into a vat under pressure, and the wine is filtered and rebottled, still under pressure. However, the *traditional method* requires a much more complicated system of sediment removal, which was invented in the Champagne region and often referred to by the French terms *rémuage* and *dégorgement*.

According to French wine law, the traditional method has to be used for making Champagne. All other French Appellation Contrôlée (AC or AP) sparkling wines also have to be made by this method (or variations on it, e.g. méthode gaillaçoise). Crémant was a term historically applied to Champagne wines that were not fully sparkling. The term may no longer be used by Champagne producers but is used in some other regions of France, e.g. Loire, Alsace, Burgundy and Limoux. The traditional method is also used for the finest-quality sparkling wines produced elsewhere in the world, most famously, Spain's Cava.

15.3 Traditional method

Although invented in England, it was in the Champagne region of northern France, that the 'traditional method' to produce a sparkling wine was perfected. Grapes are handpicked, sometimes before full ripeness, in order to produce a still wine of crisp acidity with moderate alcohol. Then, the sequence of events in the cellar is as follows.

15.3.1 Pressing
This may take place in a horizontal press or in a traditional shallow vertical 'cocquard' basket press. When pressing the grapes for Champagne, it is important not to extract colour and phenolics from the two red grape varieties (Pinot Noir and its relative Pinot Meunier) that form the backbone of most Champagnes. The traditional capacity of a basket press (and a common size for horizontal presses) is 4000 kg (4 metric tonnes) of grapes. There is a legal limit in the Champagne region as to the amount of juice that may be used for Champagne – 2500 litres per press charge of 4000 kg. This equates to 100 litres of juice per 160 kg of grapes (1 litre per 1.6 kg, which, allowing for rackings, means that nearly 1.5 kg of grapes is needed for each bottle). The first 80% of the juice extracted (2000 litres) is called the *cuvée*; the next 20% (500 litres) is called the *tailles* (tails) and is only likely to be included in the blends used for cheaper Champagnes. Any remaining juice may not be used for Champagne and is known as *vin de rebèche*.

15.3.2 Débourbage
The juice from the press is pumped into vats, which are chilled, and SO_2 and perhaps some bentonite may be added, the latter to help solids to settle. It is particularly important to ferment clear juice for sparkling wines, where the winemaker is seeking a blank canvas onto which will be developed the aromas and flavours produced during the winemaking process. Phenolics derived from the grape skins must be avoided.

15.3.3 First fermentation
In most cases, this takes place in vats – stainless steel, cement or concrete. A few Champagne houses ferment in oak barrels: e.g. Krug, Alfred Gratien and the Vintage wines of Bollinger. The fermentation is undertaken at relatively warm temperatures, 18–20°C (64.4–68°F), as a warm fermentation reduces the unwanted aromatics. The first fermentation (to an alcoholic degree of 10.5% or 11%) will take approximately 2 weeks. After this, most wines undergo a malolactic fermentation, although some houses block this if a fruitier, fresher style is required. The wines are then racked and are known as *vins clairs*.

15.3.4 Assemblage

Assemblage is a term for the blending of vins clairs and takes place in the first few months after harvest. Champagne is usually made from a blend of varieties, Pinot Meunier, Pinot Noir and Chardonnay, and the grapes grown in different parts of the region will have been vinified separately. Blending is a highly skilled and painstaking task, and a critical process in the wine's ultimate style and quality. Except in the case of Vintage Champagnes, *reserve* wines (vins clairs from previous harvests) may also be added: thus 30 or 40 different base wines may go into the blending vats. This practice of adding reserve wine is also carried out by quality-conscious producers in other regions. A Vintage Champagne is produced from grapes of a single harvest, which may have undergone selective picking and possibly just from a single vineyard site. No reserve wines are used for Vintage Champagne.

15.3.5 Addition of liqueur de tirage

Tirage is the French word for bottling. Selected cultured yeasts, able to withstand higher pressure, together with 22 or 24 g/l of cane sugar are added to the vats of blended wine to stimulate the second fermentation. Many producers prefer to use rectified concentrated must instead of sugar. The wine is then bottled in heavy sparkling wine bottles. Nowadays, these are usually sealed with crown caps containing a small plastic cup (*bidoule*) with its open end facing into the bottle. This will catch the sediment as it falls, facilitating dégorgement. However, the finest wines and those from the smaller producers are often sealed with a traditional cork (*liège*) and staple (*agrafe*).

15.3.6 Second fermentation

The bottles are taken to the cool cellars where they are laid horizontally for the second fermentation. The yeasts work on the added sugar, making an extra 1.3–1.5% alcohol and carbon dioxide, which dissolves in the wine. The gas increases the pressure in the bottle to 75–90 lbs per square inch (5–6 atm). The process may take between 14 days and 3 months, dependent upon temperature. The longer and slower the fermentation, the finer and better integrated the mousse, and the higher the resulting quality.

15.3.7 Maturation

The bottles are stored horizontally for the wine to mature, including the development of *autolytic* (yeasty) character. Ageing on the lees is a key element in the style and quality of a sparkling wine. It is another important factor that affects the wine's quality. For Champagne, the minimum legal maturation period is 15 months, or for Vintage Champagnes 3 years. NB: in the case of Spain's (basic) Cava DO, a minimum of 9 months is required. From time to time, the bottles may undergo *poignettage* (shaking and restacking), in order to prevent the yeasty sediment from sticking to the glass.

15.3.8 Rémuage

This is the process of riddling the sediment into the necks of the bottles. Traditionally, the bottles are placed into wooden easels called *pupitres*, each side of which contains 60 slanted holes, as shown in Figure 15.1.

The bottles are inserted almost horizontally, neck first. Every day for between 6 weeks and 3 months, a skilled cellar worker (rémuer) will give the bottle a twist (one eighth of a turn) and a shake, and move them slightly more vertical. At the end of the period of rémuage, the yeasty sediment will have been coaxed from the sides of the bottles into the necks. Using this system, the lighter sediment makes its way into the neck first, followed by the heavier.

Although some top Champagne houses still use rémuers, all large Champagne houses and other wineries that make sparkling wines by the traditional method now use an automated process of rémuage, which can take just 8 days. The most commonly used system is that of *gyropalettes* as shown in Figure 15.2 (NB: this photograph was taken in the Jura, in eastern France).

Figure 15.1 Pupitres. Source: Courtesy of Champagne Taittinger/Hatch Mansfield.

Figure 15.2 Gyroplattes. Source: Courtesy of Brett Jones.

These machines comprise a number of containers, each holding 504 bottles. Controlled by a computer, the machine replicates the movements of the rémuer. Some smaller houses have gyropalettes that have to be cranked by hand, and some still use a machine that looks like an automated pupitre – a *pupimatic*.

15.3.9 Stacking sur pointes

This is undertaken by some Champagne houses for all wines and by others for the best wines only. The bottles are stacked almost vertically with the neck of one bottle into the base of another for a period of up to 3 months.

15.3.10 Dégorgement

This is the disgorging of the yeasty sediment. The usual system used today is *dégorgement à la glace*. The bottles, with the sediment in the necks, are transported (still inverted), and the necks are inserted into tubs containing a freezing mixture. After approximately 6 minutes, the yeast becomes encapsulated in a slightly slushy ice pellet. The bottles are removed and turned the right way up, and a machine removes the crown cap. The pressure of the CO_2 ejects the yeast sediment. A few smaller houses still disgorge *à la volée*. The bottles are held pointing upwards at an angle of 30°, the cork or crown cap is removed manually, and the sediment is expelled into a hood.

15.3.11 Dosage (liqueur d'expedition)

The bottles are now topped up with a mixture of identical wine and a small amount of cane sugar, the precise amount varying according to the required style of wine. A few Champagne houses prefer not to do this on some wines, in order to preserve freshness. Such wines may be labelled *non dosé* or *brut nature*.

15.3.12 Corking and finishing

The final cork is now inserted and wired on. Labelling and the addition of the foil will always only be undertaken just prior to shipment, in order to avoid damage in the cellars.

15.4 Styles

Commonly used terms on the labels of many types of sparkling wines are those defining levels of sweetness (brut, demi-sec, doux) and also the degree of mousse (spumante, frizzante). Other terms used particularly in Champagne include, *Vintage*, i.e. wines produced only from a particularly good harvest, with no reserve wine added, and *non-vintage* (nv). An indication of grape variety or colour, i.e. *blanc de blancs, blanc de noirs* and *rosé*, may also be used. The term *Cuvée Prestige* is used by some producers to describe a wine made from grapes produced only in their own vineyards or perhaps from a blend made only from grand cru sites. Different production methods such as first fermentation in wood or hand disgorgement of bottles may also be carried out for Cuvée Prestige.

PART 2

Introduction to Part 2 – Wine Quality

Having gained an understanding of the processes of wine production and the numerous ways these may vary, we may now ask: how can quality in wine be defined and how can it be assessed? These apparently simple questions pose many more. Wine styles are not static – in wine regions throughout the world, the wines made today are different from those of previous generations. Few would deny that the quality of inexpensive wine is higher than ever, but the paradoxical question is: are they *better* wines? The consumer can be happy that the days of over-cropped, over-extracted and over-sulfured wines that were once commonplace are well and truly behind us. But have once distinctive regional characteristics now become blurred? And what of the so-called great wines from illustrious producers? Are they richer, softer, more voluptuous and more instantly appealing, or are they less distinctive, less a statement of place, less charming, less exciting, less enjoyable and even less drinkable than the wines of a generation or two ago? Has the drive for phenolic ripeness of fruit, which seemed unabated in the last decade of the twentieth century and first decade of the twenty-first century, led to wines that have elevated levels of alcohol and excessive fruit concentration that mask complexity?

Tastes change, and wine is, as it always has been, subject to the vagaries of fashion. In 1982, Master of Wine and Burgundy expert Anthony Hanson wrote in the first edition of his critically acclaimed book *Burgundy*: 'great Burgundy smells of shit'. If there were any raised eyebrows at the time, these were only because of Hanson's choice of language – indeed, many Burgundies had a nose of stables, farmyards and the contents thereof. By 1995, Hanson was already finding such a nose objectionable and laid the finger of blame at microbial activity. We now know that these smells have nothing to do with Pinot Noir (the variety from which pretty much all red Burgundy is made) or

Wine Production and Quality, Second Edition. Keith Grainger and Hazel Tattersall.
© 2016 John Wiley & Sons, Ltd. Published 2016 by John Wiley & Sons, Ltd.

the Burgundy terroir, but stem largely from a rogue yeast: *Brettanomyces*. Today, *Brettanomyces* is generally regarded as a fault in wine (see Chapter 21).

So, something that in 1982 was regarded by an expert taster as a sign of quality is today seen as a fault. We may draw a further example of mature Riesling. Producers in Germany and Alsace have long lauded the diesel or kerosene nose that these wines can exhibit after several years in the bottle. Many New World producers and wine critics regard such a nose as a flaw, caused by 1,1,6-trimethyl-1,2-dihydronaphthalene (TDN), also known as norisoprenoid. In common with many other wine writers, we disagree, finding such a nose part of the individual, sensuous character of this most distinctive of varieties.

There are apparent contradictions in how we assess and define wine quality. One wine can be analysed chemically and microbiologically, and be declared technically very good, yet may taste distinctly uninteresting; another wine may show technical weaknesses or even flaws, yet when tasted, it may be so full of character and true to its origin that it sends a shiver down the spine. And how can we define what constitutes a truly great wine? In his book, *The World's Greatest Wine Estates*, Robert Parker, without doubt the world's most influential wine critic, gives a workable definition of greatness as:

1 'the ability to please both the palate and the intellect';
2 'the ability to hold the taster's interest';
3 'the ability to display a singular personality';
4 'the ability to reflect the place of origin'.

It is interesting that many wine lovers bemoan the 'Parkerisation' of some wines, i.e. the sense of place is negated as producers strive to produce the style of wine driven by super-ripe fruit that they believe will earn them a high rating from the Parker team of critics.

The very concept of assessing quality in food or drink is something that does not come easily to many people. To them, a quality product is one with a 'designer' label, a well-known brand, advertised on television or in glossy magazines and which is in fashion; in other words, something that they are told is good by believable sources. The authors of this book are British. The education system, as it is now institutionalised in the United Kingdom, is bound by the straightjacket of the National Curriculum and held hostage by the need to meet targets in assessments. Consequently, it completely fails to encourage young people to develop the life-enhancing skills of discerning quality. Our colleagues in several other countries paint a not dissimilar picture. Pupils may be taught about food and nutrition but leave school unable to distinguish the fine from the mediocre, to the benefit of the many food businesses that make a lot of money from second-rate food products. It is of great concern to many wine producers that so many 'twenty-something' consumers, the so-called millennials, are not turning onto wine in the way that the previous two generations did, and most of those that do seem unwilling to leave the simplistic world of 'entry-level' quaffing liquid.

Quality may be regarded as an objective standard of excellence, with an absence of any faults. This leads us to another major issue to challenge tasters: objectivity. Objectivity is generally regarded as seeing something as it really is, uncoloured by personal preference or bias. A tasting assessment should be structured so that the taster perceives a wine to be as it truly is. However, is this achievable? The argument as to whether or not there is such thing as objectivity has induced growing perplexity during recent decades. Is objectivity the seeing of reality, something actually existing? Or is it simply taking pains to diminish or eliminate bias? Should we distinguish between ontological objectivity (seeing things as they truly are, the truth coinciding with reality) and procedural objectivity (using methods designed to eliminate personal judgements, perhaps coloured by feelings or opinions)? The key question is whether we can know if our views of reality actually correspond with it, and just how do we represent our views of reality?

In order to try and ensure rigour, validity, acceptance and realism of findings, researchers using the methods of the natural sciences use procedures that endeavour to eliminate subjectivity, which may be described as procedurally objective methods to gain an ontologically objective understanding. When tasting wines, it is important that we use techniques that are procedurally objective but realise that our assessment is not an ontologically objective one. In other words, it is a judgement, and as such it is fallible.

Judgements as to quality are, of course, framework dependent. Frameworks include those appertaining to the taster and to the wine. A taster, however open-minded he or she tries and claims to be, will work within boundaries established by training, history and culture. A Burgundian winemaker trained at the University of Dijon will have worked, lived and breathed the terroir-driven wines of the Côte d'Or. However well travelled and widely experienced they may be, they will assess a powerful fruit-driven Chardonnay from Napa Valley very differently from a UC Davis-trained American oenologist. A Muscadet de Sèvre-et-Maine Sur Lie, however well crafted and showing *tipicity de-luxe* with classic autolytic character, would, by the vast majority of trained tasters, not even be placed in the same quality league as a Montrachet Grand Cru. Yet both might be wonderfully enjoyable, just as simple cod and chips (which could accompany either wine) might please the diner as much as lemon sole prepared by the chef of a Michelin three-star restaurant. And, of course, the more illustrious the origin and producer of a wine, the higher the price, the greater the expectations of quality and the deeper the disappointment should it under perform. In other words, for both the producer and the consumer, quality is not something that can simply be bought or that can always be relied upon. Even a subtle change in any of the multifarious variables that constitute the make-up of a wine will impact positively or negatively on taste and quality. But this is just one of the reasons that makes the tasting and assessment of wines so exciting.

CHAPTER 16

Wine tasting

The history of winemaking goes back some 8000 years, which means that the history of wine tasting, at least in a basic way, is just as old. References to the taste of wine abound in works through the centuries. On 10 April 1663, the diarist Samuel Pepys wrote that he drank at the Royal Oak Tavern 'a sort of French wine called *Ho Bryen*, that had a good and most particular taste I ever met with'. Pepys's note might not have been sufficient for a pass in today's rigorous wine trade examinations, but he had the disadvantage, or should that be benefit, of not having been inundated with press releases or the pronouncements of wine writers, critics and sommeliers. He tasted the wine and gave his perceptions of it.

In 1833, *The History and Description of Modern Wines* by Cyrus Redding was published. Much of Redding's tome of over 400 pages will strike a chord with the knowledgeable wine connoisseur of today. He details the vineyard land planted, the vines, production methods, and styles and qualities of wines produced throughout Europe, South Africa and the Americas. Writing of Haut Brion, he notes that: 'The flavour resembles burning sealing wax; the bouquet savours of the violet and raspberry.' He also comments that 'the wines of Haut Brion are not bottled until six or seven years after the vintage...'. In an appendix, Redding lists 'WINES OF THE FIRST CLASS'. The list begins with Burgundy, commencing with Romanée Conti, Chambertin and Richebourg, which are described as 'the first and most delicate red wines in the world, full of rich perfume, of exquisite bouquet ...'. Next comes Gironde (Bordeaux), beginning with Lafitte [sic], Latour, Château Margaux and Haut Brion. Of Redding's other first-class wines, perhaps the inclusion of Lacryma Christi is the only one that would raise eyebrows today.

Wine Production and Quality, Second Edition. Keith Grainger and Hazel Tattersall.
© 2016 John Wiley & Sons, Ltd. Published 2016 by John Wiley & Sons, Ltd.

When, in 1920, Professor George Saintsbury's *Notes on a Cellar Book* was first published, the 75-year-old author could have had no idea that sharing his opinions of wines he had drunk over more than half a century would have such an impact upon wine lovers. Redding detailed vinous facts, as then perceived, figures, and basic tastes and perceptions. *Notes on a Cellar Book* was perhaps the beginning of a new school of art, and the precursor of a new science, of the assessment of the tastes and quality of wines, at times looking beyond simplistic descriptors. Saintsbury was to become an icon to the oenophile, having both a prestigious wine and dining club and a flagship Californian winery named in his honour. The clarets of 1888 and 1889 were, to Saintsbury, reminiscent of Browning's *A Pretty Woman* and the red wines of the south of France 'Hugonic in character'. Today's wine critic is, rightly or wrongly, usually more concerned with matters of acidity, balance, finish and perhaps the awarding of points than allusions to Browning or Hugo. Today, the professional taster also strives to be objective in the assessments made, something that Saintsbury would never claim to be.

16.1 Wine tasting and laboratory analysis

There are two basic ways by which wines may be analysed: by scientific means using laboratory equipment and by the organoleptic method, i.e. tasting. A laboratory analysis can tell us a great deal about a wine, including its alcohol by volume, the levels of free and total sulfur dioxide, total acidity, residual sugar, the amount of dissolved oxygen, and whether the wine contains disastrous spoilage compounds such as 2,4,6-trichloroanisole or 2,4,6-tribromoanisole. As we have seen, it is highly desirable that producers carry out a comprehensive laboratory analysis at various stages during the winemaking process and particularly both pre- and post-bottling. If another laboratory undertakes a duplicate analysis, the results should be replicated, allowing for any accepted margins of error. Scientific analysis can also give indications as to the wine's style, balance, flavours and quality. However, it is only by tasting a wine that we can determine these completely and accurately. If a team of trained tasters assess the same wine, they will generally reach broadly similar conclusions, although there may be dissension on some aspects, and occasionally out-and-out dispute, as we will illustrate later.

Wine is, of course, a beverage made to be drunk and (hopefully) enjoyed. The vast majority of wine produced is sold at low prices, and such wines are, at best, little more than pleasant, fruity, alcoholic drinks. As we move up the price and quality scale, wines can show remarkable diversity, individuality and inimitable characteristics of their origin. Good-quality wines excite and stimulate with their palettes of flavours and tones, their structure and complexity. Fine wine can send a shiver down the spine, fascinate, excite, move and maybe even penetrate the very soul of the taster. No amount of laboratory testing can reveal these qualities. Further, it is only by tasting that the complex intra- and interrelationship between all the components of the technical make-up of a range of wines, and human interaction with these, can be truly established. It can be argued that the perceptions of the taster are all that really matter – wine is made to be tasted not by machines but by people.

16.2 What makes a good wine taster?

Developing wine-tasting skills is not as difficult as many would imagine. While it is true that some people are born with natural talent (as with any art or craft), without practice and development such talent is wasted. People who think that they will not make good wine tasters, believing that they have a lack of inborn ability, should perhaps ask themselves some simple questions: can I see, smell and taste the difference between oranges, lemons and grapefruit, or between blackcurrants, blackberries and raspberries? If the answer is 'yes', the door is open. There are a few people, known as anosmics, who have a poor or damaged sense of smell, and obviously they are unable to become proficient tasters. A larger number of people are specific anosmics, i.e. lacking the ability to detect certain individual aromas. It is also true that some people have on the tongue a high density of fungiform papillae, and other papillae, which contain the taste buds, making them particularly sensitive to bitter sensations. It has been argued by Yale University Professor Linda Bartoshuk that this group of people are 'supertasters'. Ann Noble's group at UC Davis also established that there are no 'supertasters in general', but an individual who is a supertaster with one bitter compound, e.g. naringin(e), might be a non-taster with another, e.g. 6-*N*-propylthiouracil or caffeine. It should be noted that supertasters do not necessarily make the best wine tasters, for the intense sensations they perceive from bitterness and astringency impact on other sensations and perceptions of the balance of the wine.

With practice and concentration, the senses needed for wine tasting can be developed and refined. Memory and organisational skills also need to be developed: it is not of much use having the sensory skills to distinguish between, for example, an inexpensive young, Cabernet Sauvignon from Maule (Chile) and a fine, mature Merlot-dominated wine from Pomerol (Bordeaux, France) if one cannot organise the characteristics in the brain and remember them. Thus, the making of detailed and structured tasting notes is important – the very act of noting observations sharpens perceptions, and maintaining a consistent structure enables wines to be assessed, compared and contrasted. However, applying verbal descriptions to complex and possibly individual aroma and flavour perceptions poses many challenges. Learning, too, is important, for the taster needs to understand the reasons for the complex aromas and flavours, and to be able to describe them accurately. In short, there is no substitute for the widest-possible tasting experience, encompassing wines of all types, styles, qualities, regions and countries of origin.

When tasting wines, we are using the senses of sight, smell, taste and touch. The sense that requires the most development is that of smell. Smells create memory. You can walk into a room, and in an instant, you are reminded of another time and place – perhaps back in your infants' school classroom or in grandma's house. In the briefest of moments, your nose has detected the constituents, analysed them and passed the information to the brain, which has immediately related them to a point in the memory bank.

For most people, it is not difficult to develop the sense of smell. We live in a world in which we are conditioned to believe that many everyday smells are unpleasant, and so we try to ignore them. Walking in a city centre, we may be subjected to a mélange of traffic fumes, yesterday's takeaways and detritus of humankind, and are tempted, even programmed by the media and society, to try and ignore the onslaught. Smells may be attractive or repulsive, and an attractive smell to one person may not be to another. The smells of the human body are a key component of attraction, sexual or general, or of rejection. Animal smells in particular are offensive to many – to say that somebody smells like a dog, horse or mouse would hardly be considered a compliment!

The best way to help develop the sense of smell is to use it on every possible occasion. When walking into a room, smell it, smell the newly washed laundry, the material of clothes on a shop rail, the hedgerow blossom, even the person standing next to you. And, most importantly, commit these to memory. Expert wine tasters structure and organise a memory bank of smell and taste profiles, and thus can relate current experiences to similar ones they have encountered. Interestingly, research by Castrioto-Scanderberg *et al.* (2005), using brain monitoring by means of functional magnetic resonance imaging, shows that experienced tasters have additional areas of the brain activated during the tasting process. These are the front of the amygdala–hippocampal area activated during the actual tasting process, and the left side of the same area during the aftertaste (finish) phase.

16.3 Where and when to taste – suitable conditions

The places that wines may be tasted are perhaps as diverse as wines themselves and even less-than-technically ideal situations can have advantages. There is something magical about a tasting conducted in the vineyard, and moving from barrel to barrel in a producer's cellar can fill you with a real sense of time and place. On the other hand, exhibitions and trade shows, in spite of all the discomfort, noise and other distractions, can present a good opportunity to compare and contrast a large number of wines in a very short space of time and help place them in the context of the greater wine world.

However, for a detailed organoleptic analysis of wines, an appropriate tasting environment is required, and the ideal tasting room will have the following characteristics:

- *Large*. Plenty of room is necessary to give the taster his or her personal space and help concentrate on the tasting.
- *Light*. Good daylight is ideal, and the room (if situated in the northern hemisphere) should have large, north-facing windows. If artificial light is required, the tubes/bulbs should be colour-corrected so that the true appearance of the wines may be ascertained.
- *White tables/surfaces*. Holding the glasses over a white background is necessary to assess the appearance and show the true colour and depth of colour of the wine, uncorrupted by surrounding surfaces.
- *Free from distractions*. Extraneous noises are undesirable, and smells can severely impact on the perceived nose of the wines. Tasting rooms should not be sited near kitchens or restaurants – an amazing number of New World wineries fail to have regard for this. Tasters should avoid wearing aftershaves or perfumes, and obviously smoking should not take place in the vicinity. There is no doubt that building materials, decorations, furnishings and people all exude smells. Indeed, identical wines can be perceived differently according to the surroundings in which they are assessed.
- *Sinks and spittoons*. Spittoons, regularly emptied, are essential (see below), and sinks for emptying and rinsing glasses are desirable.

As to when to taste, the decision is unfortunately often dictated by matters beyond the taster's control. However, the ideal time is when the taster is most alert and the appetite stimulated – namely in the late morning. After a meal is certainly not the best time, for not only is the taster replete and perhaps drowsier (as all early afternoon seminar presenters know), but also the palate is jaded and confused after the tastes of the food. Also, the constituents of many foods will have an impact, positive or negative, upon taste perceptions.

16.4 Appropriate equipment

Having appropriate equipment for the tasting is most important. This includes an adequate supply of tasting glasses, water, spittoons, tasting sheets for recording notes and, at a formal sit-down event, tasting mats.

16.4.1 Tasting glasses
It is important to taste wines using appropriate glasses. Experts do not universally agree as to the detailed design of the ideal tasting glass, but certain criteria are essential:

- clear glass;
- minimum 10% crystal content;

- stem;
- fine rim;
- bowl tapering inwards towards top;
- minimum total capacity = 21 cl (approx. 7 fl oz).

Two of the key characteristics are as follows:

- *Fine rim*. A fine rim glass will roll the wine over the tip of the tongue, while an inexpensive glass with a beaded rim will throw the wine more to the centre. The tip of the tongue is the part of the mouth where we most detect sweetness, with other parts of the tongue being less sensitive to this.
- *Cup tapering inwards*. The cup of the glass must taper inwards towards the top. This will develop, concentrate and retain the nose of the wine in the headspace above the liquid, and also facilitate tilting the glass and swirling the wine. It should be noted that cut glass is not appropriate for wine tasting, as it is impossible to ascertain the true depth of colour and brightness.

Glasses manufactured to the ISO tasting glass specification (ISO 3591) remain popular among many wine tasters, both professional and amateur. A photograph of the ISO glass is shown in Figure 16.1. The ISO tasting glass is

Figure 16.1 ISO tasting glass.

Figure 16.2 Riedel Central Otago Pinot Noir glass. Source: Courtesy of Riedel.

particularly good at revealing those faults perceptible on the nose, as detailed in Chapter 21. However, the reader should be aware that some glasses marketed as ISO specification are definitely not, having such deviations as beaded rims, larger cups or inferior soda-lime manufacture. Whether ISO glasses are the best glass for tasting particular wine types is subject to much dispute. The nose of full-bodied and complex red wines certainly develops more in a larger glass having a similar basic shape; the Pinot Noir variety is more expressive in a rounder-shaped cup. Wine glass manufacturers, particularly *Riedel* and *Zwiesel*, have designs to bring out the best of individual wine types, so perhaps the only real advantage of the ISO glass is that it is a standard reference. An illustration of a Riedel Central Otago Pinot Noir glass is shown in Figure 16.2.

An appropriate tasting sample is 3–4 cl, which will be sufficient for two or three tastes. At a formal sit-down tasting of a number of wines, pouring 5 cl into the glasses gives an opportunity to return for further tastes of the wine, in order to see if there has been development in the glass and to compare and contrast with the other wines tasted. If glasses larger than the standard ISO glass are used, it is appropriate to pour correspondingly more wine.

16.4.1.1 Flutes – the ideal glasses for sparkling wines?

Tall flutes are traditionally regarded as ideal for assessing sparkling wines and without doubt are generally the best glasses for serving them. They should be fine-rimmed and preferably with a crystal content. A tasting sample comprises a quarter or third of the capacity of the flute. The quality of the mousse (sparkle) is most clearly seen and even the most delicate nose of the wine is enhanced. Interestingly, the method of manufacture of the glass makes a considerable difference to the size, consistency and longevity of the mousse in a sparkling wine. Handmade glasses give the most consistent bubbles of all, but any flute can be prepared to give a livelier mousse by, before first use, rubbing some fine glass-paper on the inside of the bottom of the cup, immediately above the stem. However, some experts believe that the complexity and particularly the expression of autolysis in a top-quality sparkling wine are best revealed in a high-quality white wine glass.

16.4.1.2 Glass washing and storage

Ideally, wine glasses should be washed by hand just in hot water. If the glasses show signs of grease or lipstick, a little detergent may be used. The glasses should be well rinsed with hot water, briefly drained then dried using a clean, dry, glass cloth that has been previously washed without the use of fabric conditioner in the washing cycle, which can leave the glass with aromas, and an oil film on the surface. Glass cloths should be changed regularly – perhaps after drying as few as six glasses. The odour of a damp or dirty glass cloth will be retained in the glass and impact on the content. At an exhibition or trade tasting where the participants take a glass from a collection on a table, the empty glass should always be nosed to check for basic cleanliness and the absence of 'off' aromas.

Glasses should not be stored bowl down on shelves, for they may pick up the smell of the shelf and develop mustiness. Obviously, standing glasses upright on shelves may lead them to collecting dust, so a rack in which glasses are held upside down by the base on pegs is perhaps ideal. In order to be sure that no taint from the glass is transmitted to the wine, it is a good idea to rinse the glass with a little of the wine to be tasted. This is also useful if tasting a number of wines from the same glass.

16.4.2 Water

There should be a supply of pure, still, mineral or spring water for the taster to refresh the palate between wines, if necessary, for drinking and perhaps rinsing glasses. The variable amount of chlorine contained in most tap water usually makes this unsuitable. Sparkling water is best avoided because the carbonic acid (H_2CO_3) content will impact upon, inter alia, assessment of the wine's acidity, any sweetness and balance. Plain, salt-free biscuits such as water biscuits may also be provided, but some tasters believe that these corrupt the palate a little. Plain, low-salt-content, soft bread is a possible alternative. Cheese, although sometimes provided at tasting events, should be avoided. The fat it contains will coat the tongue; the protein content combines with, and can soften the perception of, wine tannins; but any salt content will fight the tannins.

16.4.3 Spittoons

Spittoons, placed within easy reach of the participants, are essential at any serious wine tasting. Depending on the number of attendees and the capacity required, there are many possibilities. In the absence of purpose-made spittoons, wine-cooling buckets will suffice, perhaps with some sawdust or shredded paper in the bottom, in order to reduce splashing. There are many designs of spittoons suitable for placing on tables and larger units for standing on the floor. Consideration should be given to the construction material: plastic, stainless steel and aluminium are all good. An example of a well-designed, multi-level spittoon is shown in Figure 16.3. Unlined galvanised metal should

Figure 16.3 Multi-level spittoons.

be avoided at all cost, as wine acids can react with the galvanisation and create disgusting aromas. The importance of spitting at wine tastings cannot be overemphasised, not least because the taster needs to keep a clear head and generally avoid unnecessary ingression of alcohol. Even when wines are spat out, a tiny amount will still make its way to the stomach, and indeed a minute amount will also enter the body via the act of nosing the wines. The reader is advised not to drive after a wine-tasting event, even if all the wines have been spat out.

16.4.4 Tasting sheets

Without doubt, making notes about the wines tasted is essential. Depending on the circumstances, the notes may be brief or detailed, for personal use only or for sharing or publication. In order to facilitate note taking, tasting sheets should be prepared, listing and detailing the wines to be tasted, with space for the participants to make notes. Background and technical analysis information can also be useful, either on the tasting sheet or as a separate hand-out. A simple tasting sheet as might be used at an exhibition tasting is shown in Figure 16.4.

16.4.5 Use of tasting software

Software has been developed that facilitates the making of tasting notes on a tablet or smartphone device. The programs allow for consistently structured and detailed records to be maintained on a 'cloud' database. Organisers of major trade tastings can submit details of the wines to be tasted to the software provider, who makes the information available for download to a mobile device.

16.4.6 Tasting mats

If a number of wines are to be assessed at a formal sit-down tasting, each wine should have its own glass, placed on a paper tasting mat printed with circles of a size similar to the bases of the tasting glasses, each circle numbered and corresponding to the listed order of the wines on the tasting sheets. A simple tasting mat is shown in Figure 16.5.

16.5 Tasting order

If there are many wines to be tasted, of varying styles and qualities, it is sensible to do so in a considered order. There are several guidelines, but unfortunately many of these conflict:

- dry white wines should be tasted before sweet;
- light-bodied wines should be tasted before full-bodied;
- wines light in tannin should be tasted before those with high levels of tannin;

BORDEAUX – QUALITY, DIVERSITY & VALUE
BROADWAY WINE SOCIETY 24TH JUNE 2015
PRESENTER: KEITH GRAINGER

	WINE	AOP	BLEND	STOCKIST	TASTING NOTES
DRY WHITE WINES					
1	Château Tour Léognan 2011	Pessac-Léognan	70% Sauvignon Blanc 30% Sémillon	Waitrose £14.99	
RED WINES					
2	Château La Ramonette 2010	Bordeaux	80% Merlot 10% Cabernet Sauvignon 10% Cabernet Franc	Broadway Wine Company £7.75	
3	Château Pey La Tour Reserve du Château 2010	Bordeaux Supérieur	90% Merlot 5% Cabernet Sauvignon 3% Cabernet Franc 2% Petit Verdot	The Wine Society £10.59	
4	Château Segonzac 2011 (Oak Aged Cuvée)	Blaye Côtes de Bordeaux	60% Merlot 20% Cabernet Sauvignon 20% Malbec	Waitrose £10.29	
5	Château Fonguillon 2012	Montagne Saint-Emilion	70% Merlot 15% Cabernet Franc 12% Cabernet Sauvignon 3% Malbec	Tesco £10.49	
6	Château Le Vieux Fort 2010	Médoc Cru Bourgeois	55% Cabernet Sauvignon 40% Merlot 5% Petit Verdot	Waitrose £13.49/ £9.99	
7	La Terrasse de la Garde 2010	Pessac-Léognan	60% Merlot 38% Cabernet Sauvignon 2% Cabernet Franc	Sainsbury's £14.99	
SWEET WINE					
8	Château Liot 2010	Sauternes	90% Sémillon 10% Sauvignon Blanc	Waitrose £16.99 37.5cL	

www.bordeaux.com/uk

Figure 16.4 Simple tasting sheet.

- young wines should usually be tasted before old;
- modest-quality wines should be tasted before those of high quality.

Whether white wines should be tasted before red, and sparkling before still is now a matter of some dispute among tasters. It will easily be seen that trying

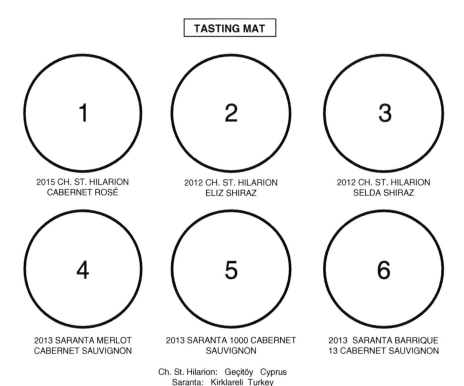

Figure 16.5 Tasting mat.

to sort a sensible tasting order for a wide range of wine styles and qualities proves challenging, especially as the characteristics of each wine may not be as anticipated. At a trade exhibition, tasting well in excess of 100 wines is not uncommon, and even the most experienced tasters can suffer fatigue. It can be particularly difficult to taste a large number of sparkling wines, as the high acidity numbs the palate. Also challenging to a taster is a large quantity of red wines that are high in palate-numbing tannins. Many Master of Wine (MW) tasters advocate tasting red before white. The acid attack of whites makes red tannins seem more aggressive. The aromas of white wines are easy to assess after red wines.

16.6 Temperature of wines for tasting

The temperature of wines presented for tasting analysis is not necessarily that at which we would wish to drink them. This is particularly true of white wines, which many people prefer to drink relatively cold, perhaps 8–12°C

(46–54°F). Coldness numbs the palate, and white, rosé and sparkling wines are best tasted cool (12–15°C) rather than cold. Conversely, many people drink red wines at a relatively warm temperature. The expression 'room temperature' does not mean 22°C or more, the temperature of many a living room. The French language has an expression chambré, which refers to bringing wines from cellar (storage) to serving temperature. La chambre is, of course, the bedroom, which will be cooler than the living or dining rooms.

Red wines are best tasted at 16–18°C, with the lighter reds, including those made from Pinot Noir, at the lower end of the scale. Some consumers might prefer to drink full-bodied reds a degree or two warmer than this, but tepid wines are distinctly unappetising.

16.7 Tasting for specific purposes

The way a tasting is approached and the type and detail of notes made may vary according to the purpose of the tasting and agenda of the taster. A supermarket or merchant buyer needs to consider marketability, consumer preferences, how a wine relates to others on the list and price point. An independent merchant selling fine wines may pay high regard to how true a wine is to its origin, often referred to in the wine world as *typicity* and, in the case of young wines, the ageing potential indicated by the tasting. A restaurateur particularly needs to have in mind to match the wine with dishes on the menu. A winemaker choosing and preparing a blend looks beyond the taste of the component wine in the glass to the contribution it might make, in variable amounts, to a finished blend. This said, it is important for the taster to assess each wine as completely and objectively as possible by adopting a consistent and structured tasting technique as detailed below.

16.8 Structured tasting technique

Most people do not really taste wine; they simply drink it. But by undertaking a detailed and considered tasting ritual, all that the wine has to offer, good and bad, is assessed. Four headings should be considered in tasting each wine: appearance, nose, palate and conclusions. We examine our approach to these very briefly here, and in more detail in Chapters 17–20.

16.8.1 Appearance
The appearance can tell much about a wine and give indications as to origin, style, quality and maturity, as well as revealing some possible faults. The appearance of the wine should be examined in several ways, particularly by holding the glass at an angle of approximately 30° from the horizontal over a

Figure 16.6 Wine glass tilted to assess appearance.

white background – perhaps a tablecloth or sheet of white paper. Such an assessment is shown in Figure 16.6. This will enable the clarity, the intensity of colour and the true colour of the wine to be seen, uncorrupted by other colours in the room. Looking straight down on a glass of wine that is standing on a white background is also useful for determining the intensity of colour. Other observations should be made: e.g. any legs or tears that run down the surface of the glass, which are best seen by holding the glass at eye level and giving a gentle swirl to coat the wall of the cup.

16.8.2 Nose
A wine should be given a short sniff to check its condition. Most faults will be revealed upon nosing, and if the wine is clearly out of condition, we probably will not wish to continue past this point. Also, some of the more delicate volatile compounds are most easily ascertained by a short, gentle sniff or two. Further, the longer we are exposed to the aromas, the less sensitive we become to them, so these first, gentle sniffs are important. Assuming that no faults have been revealed on this initial nosing, the wine should now be aerated in order to help release the volatile aromas – i.e. putting air into the wine

will make the nose more pronounced. The usual way of achieving this is by swirling the wine round the glass several times. This skill is quickly learned, but should you experience difficulty, the glass can be steadied on the side of a loosely clenched fist as you swirl. If the wine seems very closed, i.e. showing few aromas, the glass, appropriately covered, can be shaken vigorously for a second or two. This is seldom necessary other than for the poorest of wines, and subsequently the wine should be given several short sniffs. Try placing the nose at various points of the glass to see if the aromas are more pronounced or different. Very long sniffs should be avoided at this stage, too, on account of the numbing effect. We should note the intensity of the nose – put very simply, how much smell do we get from the wine? The development of the nose, explained in detail in Chapter 18, will indicate the present stage of the wine in its maturity cycle. Most importantly, we should analyse the aroma characteristics.

16.8.3 Palate

A novice watching an accomplished and experienced taster at work will perhaps be unsure as to whether to view the ritual with laughter, derision or wonder. Observing the taster nose the wine in detail may already raise eyebrows, but watching the subsequent ritual of slurping and chewing the liquid may seem like overacted theatre. However, a simple exercise will convince even the most sceptical of the value of a professional approach. A small sample of wine should be poured into a glass, drunk as one would normally drink and then reflected on for a moment. Then, another mouthful, 1 or 1.5 cl is an appropriate amount, should be taken and assessed using the professional approach. It is important to take a suitable quantity of wine so it is not overdiluted by saliva, and there is sufficient to assess it fully. The wine should be rolled over the tip of the tongue and air should be breathed into the liquid. This is not difficult: the lips should be pursed with the head forward as air is drawn into the wine. The taster should not be concerned about the slurping noises made during this operation and forget the childhood scolding given by mother! Now, the liquid should be thrown around the mouth – over the tongue, gums and teeth, and into the cheeks. The wine should be chewed, making sure that the sides and back of the tongue are covered, and a little more air taken in. It will take 20 s or so to give a thorough assessment, but there is no point in retaining the wine in the mouth for longer than this, as it will be diluted by saliva, and the palate will have become numbed. Finally, the wine should be spat out, and the taster should breathe out slowly and reflect. The huge number of sensations experienced during this exercise compared with simply drinking the wine will astound. The purpose of breathing air into the wine is to facilitate the vaporisation of the volatile compounds that travel via the retro-nasal passage to be sensed by the olfactory bulb.

The tip of the tongue will detect the level of sweetness, and the sides of the tongue and cheeks the acidity. Other areas of the tongue also detect these sensations, and this topic is discussed in Chapter 19. Tannins, normally only really important when tasting red wines, will be sensed particularly on the teeth and gums. The level of alcohol is felt as a warming sensation, especially on (but not limited to) the back of the mouth. The body of the wine is the weight of wine in the mouth. When tasting sparkling wines, the entire mouth will also feel the sensations of the mousse. For the flavour intensity and flavour characteristics, it is not just the mouth at work but also the olfactory bulb that will receive volatile compounds via the retro-nasal passage. Finally, the all-important finish of the wine is assessed – this is the amount of time the flavours are retained after it has been spat out.

16.8.4 Conclusions

Having thoroughly assessed the wine, judgements and conclusions may now be made. Perhaps the most important consideration is the quality. Assuming one is tasting wines already on the market, the price or at least the price category should be determined. Of course, the relationship between quality and price is the key to assessing value. The readiness for drinking should also be decided upon. If the wine is being tasted totally blind, that is without the taster knowing previously what the wine is, now is the time to mentally collate the information obtained during the structured tasting and reach a conclusion on key points. These might include the grape variety or varieties, vintage, state of maturity and the country, region, district, commune or even individual property of origin.

16.9 The importance of keeping notes

Making, organising and keeping structured tasting notes is essential to improving the tasting technique, to enable wines tasted over time to be compared and contrasted, and for providing a source of reference. The amount of detail included in the notes will obviously depend on the circumstances of the tasting, the time available and the taster's specific focus and requirements. It is important to avoid any possible subsequent ambiguity or misunderstanding. This is vital if the notes are not intended for the taster's private use or are to be made for later publication.

The following chapters cover in more detail the structured tasting technique and the headings under which we consider the wines and make notes, and also detail some appropriate descriptors. The tasting structure and example tasting terms used herein are generally those of the Systematic Approach to Tasting of the Diploma Level of the Wine & Spirit Education Trust (see Appendix).

There are, of course, many other tasting expressions that the taster will wish to use, and the terms detailed in the ranges that follow are far from exhaustive. However, unless the notes are purely for their own use, the taster is cautioned against using terms that are particularly personal to him or her. A note that the nose of the wine is 'reminiscent of Aunt Edna's lounge' will mean nothing to the reader who has not visited that particular room, which perhaps exudes the aromas of lilies, potpourri, wax polish and Earl Grey tea. The retention and subsequent review of at least a selection of one's tasting notes, perhaps transcribed into notebooks or inputted into a computer, not only help to develop and refine technique but also provide a source of reference of the aromas and flavours of the wines assessed.

CHAPTER 17

Appearance

To many general wine drinkers and novice tasters, the detailed examination by a professional of a wine's appearance seems something of a pointless exercise. However, the appearance reveals much about a wine, giving indications as to style, quality and maturity, and may reveal some possible faults. In this chapter, we examine the assessment of the appearance of a wine and consider the following:

- clarity and brightness;
- intensity;
- colour;
- other observations.

17.1 Clarity and brightness

A wine that has finished fermentation and has been stabilised for sale should be clear. Brightness is a sign of healthiness. Brightness describes how light is reflected off the surface of a wine. This is related to the pH – the lower the pH, the brighter the wine usually is. Clarity describes how light is scattered as it passes through the body of the wine, and this is related to turbidity, i.e. suspended particles. If a mature wine that has thrown sediment has been carelessly handled, there may be fine or larger particles in suspension, but any finished wine that appears cloudy, oily, milky or otherwise murky is probably faulty. Possible faults are detailed in Chapter 21.

We may consider the clarity and brightness of a wine on the following simple scale:

clear – hazy/bright – dull (faulty?)

Wine Production and Quality, Second Edition. Keith Grainger and Hazel Tattersall.

Wines from the New World are often brighter than those from Europe. Wines that have high acidity, particularly if tartaric acid has been added in the winemaking process, may appear to be especially bright. A young wine that appears dull probably has a high pH (low acidity), which is indicative of poor quality and a total lack of ageing potential. As wines mature, they lose brightness, and with over-maturity they become dull. A decrepit wine will look very flat.

17.2 Intensity

The intensity, i.e. the depth of colour, should now be noted. An impression of the depth of colour can be obtained by looking directly down onto the wine in a glass that is standing on a white background. Figure 17.1 shows a 6-year-old Amarone della Valpolicella viewed from above. A more detailed examination is obtained by tilting the glass, again over a white background, and looking particularly at the heart of the wine. Figure 17.2 shows the same Amarone viewed in this way. It is important when comparing and contrasting several wines to ensure that the glasses contain a similar quantity of liquid.

Figure 17.1 Looking down for an impression of intensity.

Figure 17.2 The same wines as Figure 17.1, glass tilted to 30°.

We may consider the intensity of a wine on the following scale:

pale – medium – deep

The taster may wish to add additional simple descriptors, e.g. extremely deep. Generally speaking, wines that are pale in intensity will be lighter in flavour and body than those that are deep, but this is not always the case. Whites that have had extensive barrel ageing will have a greater intensity. Some red grape varieties such as Merlot and Syrah (also known as Shiraz) are deep coloured, especially when from a hot climate and from low-yielding vines. Conversely, several red grape varieties including Pinot Noir and, to a lesser extent, Nebbiolo produce wines that are usually not very deep coloured, although there are some notable exceptions to this, e.g. some Pinot Noirs from New Zealand's Central Otago region. However, paleness in a red wine, especially a young wine, is generally indicative of a lack of concentration, perhaps as a result of high-yielding vines, cool climate, difficult vintage or inadequate extraction. Full-bodied and concentrated reds, especially those from hot climates, will be deep or opaque in youth, and this suggests intense flavours and may be indicative of good quality. However, occasionally, extremely deep-coloured red wines can be surprisingly light in flavour. There are a few red varieties that have coloured flesh as well as black skins (e.g. Alicante Bouschet) but are surprisingly light in flavour, and these *teinturier* grapes may (very occasionally) be used in a blend to deepen the colour. Further, very deep colours can sometimes indicate winemaking methods that have focused on colour, rather than flavour extraction. Mindful of the association in the eyes of drinkers (and critics) of deep red wines and perceived high quality, winemakers can manipulate intensity, for example, by the addition of *8000 color* or *megapurple*, concentrates made from grape skins and seeds. Thus, the taster is cautioned against drawing conclusions at this stage.

17.3 Colour

Many factors affect the colour of a wine, including climate and region of production, grape variety or varieties, grape ripeness, vinification techniques including any barrel ageing and maturity.

17.3.1 White wines

White wines can vary in colour from almost water clear to deep gold or even amber. We may consider the colour of a white wine on the following scale:

lemon green – lemon – gold – amber – brown

Again, the taster may wish to add their own descriptors. Some white varieties such as Sauvignon Blanc generally produce wines at the lemon-green end of the range, while others such as Gewürztraminer (which has a more heavily pigmented skin) give straw or gold colours. Whites from cooler climates often appear lemon green or lemon, and those from warmer areas gold. Sweeter wines such as Sauternes are usually gold, even in youth, but all whites will darken with age. The rate at which this takes place varies considerably depending on several factors, particularly pH and acidity (higher acidity slows down the process) and how well the bottle has been stored. In 2012, the authors tasted a delightful Wehlener Sonnenuhr Riesling Kabinett (S.A. Prüm) from the Mosel region in Germany and from the less-than-highly regarded 1987 vintage. At 25 years old, the wine was very bright and the colour still lemon green, having been preserved by a little CO_2 from the fermentation remaining in the wine and particularly the high acidity, the result of having been produced not only in a cool region but also in a cool year.

17.3.2 Rosé wines

Of all wine types, rosé is the category that is made to appeal on the eye. Winemakers and marketing departments know that an appealing appearance is crucial to the drinker's perceptions of style and quality. The colour of rosé

Figure 17.3 A range of rosé wines.

wines depends on several factors, particularly the production techniques, e.g. whether the colour is the result of skin contact in the press, or if the wine has been made by the *saignée* method, in which juice from a vat of fermenting, crushed red grapes is extracted after 6–24 h or so. We may consider the colour of a rosé wine on the following scale:

pink – salmon – orange – onion skin

There is a very wide range of possible colours, and also of intensity. Any orange tint can be a danger sign, indicating oxidation and that the wine is beginning to develop bitter phenolics, which will be perceived as such on the palate. An illustration of a range of rosé wines is shown in Figure 17.3.

17.3.3 Red wines

Red wines can vary in colour from purple to mahogany, losing colour with age. Purple is indicative of a young red wine, some being so intensely coloured that they appear almost to be blue purple. As the wine begins to age, the purple tones lighten to ruby, with further age to a warm brick red colour, and with full maturity possibly to garnet or even tawny. Figure 17.4 illustrates the appearance of a young (under 5 years old) Cru Bourgeois from

Figure 17.4 A young Cru Bourgeois from Bordeaux.

Bordeaux, made from Cabernet Sauvignon and Merlot. Figure 17.5 shows a very mature Cru Classé Bordeaux, some 45 years old.

We may consider the colour of a red wine on the following scale:

purple – ruby – garnet – tawny – brown

Red wines that have undergone lengthy barrel ageing, in which a controlled oxygenation has been taking place, lose colour faster than those bottled early and just aged reductively in bottle. A good illustration of this is the contrast between a 10-year-old Tawny Port, which has been matured in pipe (cask), and a Vintage Port (not Late Bottled Vintage – LBV) about 12 years old, which has been aged reductively in bottle for all but the first 2 years or so. A wine that is brown in colour is tired, oxidised and probably undrinkable. Red and white wines that are totally oxidised are pretty much indistinguishable in colour.

Figure 17.5 A very mature Cru Classé Bordeaux, some 45 years old.

17.3.4 Rim/core

The gradation of colour from the heart or core of the wine to the rim should be noted. Of course, the greatest intensity of colour is at the heart, but in the area approaching the rim, not only will the colour be paler, but it will change. For example, a wine that is ruby coloured at its core may gradate to brick red or garnet tones towards the rim, indicating maturity. As the rim of the wine touches the glass, the last millimetre or two will be water clear. The distance of the gradation in colour will vary from just a couple of millimetres in a young wine to perhaps a centimetre or more in a mature example. Mature white wines, too, will have considerable gradation in colour approaching the final few millimetres of the rim, which again will be water clear. The colour of the rim should be noted, and the width should be considered and noted on the following scale:

broad – medium – narrow

If we contrast Figure 17.4 with Figure 17.5, there is a striking difference in colour at the heart of the wines and dramatic changes to the broad rim on the mature wine, paling to mahogany as it laps the glass.

17.4 Other observations

These may include, inter alia, legs/tears, deposit, petillance and bubbles.

17.4.1 Bubbles
Obviously, bubbles are a key feature of sparkling wines, but a small amount may also be observed on still wines.

17.4.1.1 Still wines
Occasionally, the presence of bubbles on a still wine could be indicative of a fault – either an alcoholic or malolactic fermentation is taking place or has taken place in the bottle (see Chapter 21). However, still wines in good condition may contain bubbles. Gases, namely, CO_2, N_2 and Ar, may be used as a blanket at various stages of winemaking in order to prevent oxidation or other spoilage. If a very fresh style of wine is being produced, it is common to flush the vats and blanket the wine with one of these gases at bottling time. Bottles, too, may be pre-evacuated of oxygen and gas-sparged immediately prior to filling. Some of the gas (particularly CO_2) may become dissolved in the wine – this does not generally detract from quality and can often add a sensation of freshness. Some wines, e.g. wines from the Mosel region of Germany, may naturally retain some CO_2 from the alcoholic fermentation. In the case of still wines, a brief observation of the size and quantity of bubbles should be made. The bubbles may just appear on the tasting glass, in which case they are likely to be large, on the rim or in the heart of the wine.

17.4.1.2 Sparkling wines
The quality of the mousse is considered to be an essential part of the overall quality of sparkling wines. Notes should be made on the size, quantity and consistency of the bubbles. The bubbles may rise from the base of the cup or from a seemingly random point in the heart of the wine. Generally speaking, small bubbles are indicative of a desirable cool, slow, second fermentation, especially one that has taken place in the bottle as in Champagne and other high-quality sparkling wines made by the traditional method. It should be noted that bubbles can vary somewhat according to the type and washing of the tasting glass; see Chapter 16, Section 16.4.1. Figure 17.6 shows the fine, continuous and even bubbles, with pearl-like strings, of a good-quality Champagne, with a medium intensity of colour.

Figure 17.6 Fine, pearl-like bubbles of a good Champagne.

Figure 17.7 shows fewer and inconsistent bubbles in an inexpensive sparkling wine (a very pale Sauvignon Blanc) made by the Charmat method.

17.4.2 Legs

One of the most misunderstood visual aspects of wine tasting is the presence or otherwise of legs, often referred to as tears. The wine should be swirled in the glass, held to eye level, waiting for several seconds, viewed horizontally and observed as to how the swirled wine runs back down the glass. If the liquid congeals into little tears, arches or rivers running down the glass, these are called legs. The legs may be broad or narrow (thin), short or long, and run slowly or more quickly down the glass. Notes made should describe them as such, and subjective terms such as 'good legs' are to be avoided. Wines that contain a high degree of alcohol, will normally show broad legs, formed by the difference in surface tension between water and alcohol, and the differential evaporation of alcohol, influenced by sugar and glycol. Several authors and critics claim that legs are purely a sign of high alcohol, glycerol or very

Figure 17.7 Inconsistent bubbles in a Charmat method sparkling wine.

high residual sugar, but this is refuted by examples of their presence in high-quality wines that are relatively light in alcohol, for example, fine Riesling Kabinetts from the Mosel region in Germany. The amount of dry extract also contributes to legs. One of the most 'leggy' wines ever assessed by the authors was a dry white 1998 Château Miravel, Côtes de Provence, made from the Rolle (Vermentino) variety, with a low yield of just 30 hl/ha and with an alcohol content of just 11.5%.

The amount and type of legs, if any, can be very much dependent on the condition, and particularly the washing and drying of the tasting glass. The same wine poured into several apparently identical glasses, but washed at different times, may show striking differences in the legs.

17.4.3 Deposits
Any deposit in the glass should be noted. These may comprise tannin sediments (which are simply the coagulation of phenolic substances) in the case of red wines or tartrate crystals in either reds or whites. Tartrate crystals are *not*

a wine fault and are often visible in wines of the very highest quality. Thick deposits in red wines are usually tannin-stained tartrate deposits (especially in low pH wines). In white wines, however, they can look alarmingly like pieces of broken glass and really worry the consumer, but they are completely harmless. The crystals are most likely a precipitate of potassium bitartrate ($KC_4H_5O_6$) or occasionally calcium tartrate ($C_4H_4CaO_6$). They are often found in bottles of German wine, which tend to be high in tartaric acid. They may precipitate if the wine gets very cool – perhaps in a cold cellar or refrigerator. Many winemakers go to great lengths to try to ensure that the crystals do not appear in the bottle, and the reader is referred to Chapter 13, Section 13.3. The money and time spent by the industry on such treatments might be better invested in consumer wine education.

CHAPTER 18

Nose

In this chapter, we cover the characteristics of a wine that may be identified by nosing. We also discuss the origins or sources of these and examine the concept of primary, secondary and tertiary aromas. We also illustrate words that may be used as descriptors for the various aromas perceived.

The olfactory epithelium, situated at the top of the nasal cavity, is a very sensitive organ. As we will see in Chapter 19, the tongue reveals only a very limited number of tastes, and most of the 'taste' sensations are detected by the receptor cells of the olfactory epithelium, received via either the nasal or retro-nasal passage. The information is turned into electrical signals and sent via the olfactory bulb to the olfactory cortex in the brain. We will discuss the sensations transmitted via the retro-nasal passage in Chapter 19. There is no doubt that repeated and overexposure to particular smells reduces sensitivity to them, and this can be an issue for winemakers who are regularly and at times continuously exposed to odours such as SO_2.

When nosing a wine, we are, of course, smelling the air space in the glass, above the surface of the liquid. It is important that there is plenty of head space for the aromas, the volatile compounds in the wine, to develop. An inwards tapering bowl of a well-designed tasting glass enables aromas to be retained in the headspace. It is worth noting that when pouring wines for general serving, the glasses should be filled to no more than 50% of their capacity in order for the nose to be appreciated. It is unfortunate that even in bars that offer an interesting range of wines by the glass, they are often filled close to the brim to comply with legal sales measures.

The nose of the wine should be assessed in the following stages:

- condition;
- intensity;

Wine Production and Quality, Second Edition. Keith Grainger and Hazel Tattersall.
© 2016 John Wiley & Sons, Ltd. Published 2016 by John Wiley & Sons, Ltd.

- aroma characteristics;
- development.

As we will see below, the wine should be first nosed without swirling, then swirled around the glass to vaporise the volatile compounds and given a comprehensive nosing.

18.1 Condition

The initial nosing of the wine will assess the condition: many possible faults are apparent at this stage. The wine should not be swirled prior to this, and just one or two short sniffs are all that is required. Basically, we are checking if we want to go any further in the tasting procedure. For example, if the wine smells of damp sack or a damp musty cellar, vinegar, struck matches or old cream Sherry, it is faulty, and the nose may be described as unclean. Chapter 21 details faults and which of these are detectable upon the nose. Depending on the nature and severity of any fault revealed, a decision has to be made as to whether or not to proceed with the tasting of the wine concerned. A nose that is free from faults is described as clean.

A note should be made on the condition of the nose:

clean – unclean + fault specified

18.2 Intensity

This is simply assessing how strong or 'loud' the nose of the wine is and can give an indication of quality. A nose light in intensity may be expected from a simple, inexpensive wine, with a more pronounced nose indicative of a higher quality. However, high-quality reds in particular can be very closed in youth. Conversely, there are some grape varieties that nearly always give a very intense nose even in a wine of modest quality, especially aromatic whites. These include members of the Muscat family, Gewürztraminer and Argentina's three Torrontés varieties.

We may consider and note intensity on the following scale:

light – medium (–) – medium – medium (+) – pronounced

18.3 Development

It is perhaps convenient if we consider development before aroma characteristics, although when making a tasting note, the aroma characteristics might be noted first.

Development is assessing on the nose the state of maturity of the wine. Wines have a lifespan depending on many factors, and maturity should not be confused with age. Generally speaking, the higher the quality of the wine, the longer is the lifespan. The finest-quality reds, including many cru classés from Bordeaux, or classic wines from the northern Rhône, may require 10 years or more, even to approach their peak. Conversely, a simple branded Bordeaux or Côtes du Rhône may be past its best at an age of 3 or 4 years.

To understand the concept of nose development, we need to consider the basic sources of wine aromas. These may be classified into three main groups:

- primary;
- secondary;
- tertiary.

18.3.1 Primary aromas

Primary aromas are those derived from the grapes, and are fruity or floral in character. These normally indicate a wine that is in a youthful stage of development. Varietal aromas usually belong in this group. The blackcurrant nose of a young Cabernet Sauvignon, or the elderflower nose of some young Sauvignon Blanc or Seyval Blanc wines would fall into this group.

18.3.2 Secondary aromas

Secondary aromas are the 'vinous' aromas resulting from the fermentation – put simply, the smells that are different in wine to those of unfermented grape juice. Numerous esters are generated in the fermentation, and these are often assertive on the nose of young wines, imparting pear, banana or even boiled sweet or bubble gum characteristics. If present, the by-products of MLF, bâtonnage and oak extracts are included in the secondary aroma group. Also included are aromas of butter, cream, vanilla, coconut, nuts and toast.

18.3.3 Tertiary aromas

Tertiary aromas are the result of the ageing process of the wine in tank, barrel and/or bottle. During this time, there will be many chemical reactions. The wine will already contain some dissolved oxygen, and for wines that are barrel matured, further oxygen may be absorbed through the cask. Provided the barrels are kept topped up, there will only be a small amount of beneficial oxygenation taking place, which will, inter alia, increase the aldehyde content of the wine. Wine matured in new barrels will also absorb oak compounds including vanillin, lignin and tannin. Those matured in second and third fill barrels will also do so, but to a lesser extent. Wines that are not matured in barrel may be micro-oxygenated and 'oaked' in other ways, and this was discussed in Chapter 12.

Maturation in bottle results in changes to the volatile compounds of the wine. It is generally accepted that bottle maturation is of a reductive nature, that is, the changes take place anaerobically, and the oxygen content of the wine is reduced. It is this process that makes bottle ageing a necessity for the maturation of many fine wines, especially full-bodied reds. A wine that does not have the capacity to reduce is a dead wine. However, *reductivity* is a fault, and this will be discussed in Chapter 21. As good wines mature, the tertiary aromas develop, and these can exhibit a complex array of seamless, interwoven characteristics, at best all in total harmony. Tertiary aromas are those that often evoke the most descriptive comments: saddles, shoe-leather, Havana cigar, woodland floor, autumnal gardens – the list is almost endless. Many vegetal characteristics such as cabbage-like smells also belong to the tertiary group. All such observations should be noted when we note the details of the wine's aroma.

When wines are aged beyond their projected lifespan, or suffer premature decline (perhaps owing to poor storage), all the primary and many of the secondary aromas will be lost. Of the tertiary aromas that remain, some such as caramel, toffee or soaked fruitcake may be pleasant, but many will not. The smell of rotting cabbage, stale sweat, old trainers and burnt saucepans may come to the fore, and every hint of complexity will have been drained out.

We may consider and note development of the nose on the following scale:

youthful – developing – fully developed – tired/past its best

It should be noted that some wines go from youthful to tired without ever passing through the fully developed stages. Examples include most Beaujolais and inexpensive types of Valpolicella. However, there are also types of wine that are bottled fully developed, such as Sherry, Tawny Port and some sparkling wines.

18.4 Aroma characteristics

It should be noted at this point that many tasters and wine writers distinguish between the terms 'aroma' and 'bouquet'. They use the term 'aroma' when referring to the primary aromas derived from the grapes. 'Bouquet' would encompass the assembled characteristics resulting from changes that have taken place during fermentation (secondary aromas), and particularly maturation (tertiary aromas). As we will see, the distinction is not clear cut, and we will use the word 'aroma' as an all-encompassing term.

Over 400 wine odour compounds have been identified. Their concentrations vary from <100 ng to 300 mg/l, and the olfactory perception thresholds of the compounds vary from 50 pg to 100 mg/l. The compounds in grapes that are precursors of wine flavours include free amino acids, phospholipids,

Table 18.1 WSET® diploma systematic approach to wine tasting

Describing aroma and flavour characteristics (extracted)
Primary aroma/flavour clusters: the flavours of the grape
Floral
Acacia, honeysuckle, chamomile, elderflower, geranium, blossom, rose, violet, iris
Green fruit
Green apple, red apple, gooseberry, pear, peardrop, custard apple, quince, grape
Citrus fruit
Grapefruit, lemon, lime (juice or zest?), orange peel, lemon peel
Stone fruit
Peach, apricot, nectarine
Tropical fruit
Banana, lychee, mango, melon, passion fruit, pineapple
Red fruit
Redcurrant, cranberry, raspberry, strawberry, red cherry, red plum
Black fruit
Blackcurrant, blackberry, bramble, blueberry, black cherry, black plum
Dried fruit
Fig, prune, raisin, sultana, kirsch, preserved fruits
Herbaceous
Green bell pepper (capsicum) grass, tomato leaf, asparagus, blackcurrant leaf
Herbal
Eucalyptus, mint, medicinal, lavender, fennel, dill
Pungent spice
Black/white pepper, liquorice, juniper

Secondary aroma/flavour clusters: the flavours of winemaking
Yeast
Biscuit, bread, toast, pastry, brioche, bread dough, cheese, yoghurt
MLF
Butter, cheese, cream, yoghurt
Oak
Vanilla, cloves, nutmeg, coconut, butterscotch, toast, cedar, charred wood, smoke, resinous
Other
Smoke, coffee, flint, wet stones, wet wool, rubber

Tertiary aroma/flavour clusters: the flavours of wine
Deliberate oxidation
Almond, marzipan, coconut, hazelnut, walnut, chocolate, coffee, toffee, caramel
Fruit development (white)
Dried apricot, marmalade, dried apple, dried banana, etc.
Fruit development (red)
Fig, prune, tar, cooked blackberry, cooked black cherry, cooked strawberry, etc.
Bottle age (white)
Petrol, kerosene, cinnamon, ginger, nutmeg, toast, nutty, cereal, mushroom, hay, honey
Bottle age (red)
Leather, forest floor, earth, mushroom, game, cedar, tobacco, vegetal, wet leaves, savoury, meaty, farmyard

glycolipids, aldehydes and phenols. Alkyl esters, a result of fermentation, are important compounds that give secondary aroma characteristics. Terpenes, present in grapes, are unchanged by the fermentation process and contribute to primary aromas. However, maturation and ageing of the wine may result in their being changed and contributing to tertiary aromas. Young wines made from grapes that have a high terpene content such as the Muscat family, Gewürztraminer and Riesling can have a nose that screams of primary fruit and show overt grape-like aromas. Many compounds that give varietal aromas remain largely unchanged by the fermentation process, and varietal aromas, e.g. the pronounced blackcurrant or cassis aromas of Cabernet Sauvignon, are considered as primary.

There has been considerable research over the last couple of decades into the compounds that contribute to wine aromas, particularly the varietal aromas. By way of example, the aromas commonly found in young wines made from Sauvignon Blanc, and the compounds that result in such aromas, are as follows:

• boxwood – broom: 4-mercapto-4-methylpentan-2-one;
• citrus peel: 4-mercapto-4-methylpentan-2-ol;
• grapefruit, passion fruit: 3-meracptohexanol;
• boxwood – broom: 3-mercaptohexylacetate;
• green pepper, grass: 2-methoxy-3-isobutyl pyrazine.

We may consider wine aromas in six basic groups:

fruits – flowers – spices – vegetables – oak aromas – other

Each of these basic groups comprises several subgroups, which in turn contain individual aromas and, when we come to taste the wine, flavours. Table 18.1, a lexicon, shows the aroma and flavour sources, subgroups and individual aromas/flavours. It should be noted that the descriptors in this table are far from exhaustive, and some aromas may have more than one source.

All aromas sensed should be noted, and when detailing individual terms these may be linked to known varietal characters. For example, green apple, lime, peach and mango are just some of the aromas that may be associated with Riesling; strawberry, raspberry, red cherries, green leaf and mushroom are typical aromas associated with Pinot Noir. Any oak aromas including vanilla, toast, smoke, nuts and coconut should particularly be noted.

CHAPTER 19

Palate

Palate is a convenient expression to describe the taste of wine once it has entered the mouth. In this chapter, we assess the taste and tactile sensations detected in the mouth, particularly on the tongue, and the flavour characteristics detected as a result of the wine's volatile compounds being breathed through the retro-nasal passage at the back of the mouth and transmitted to the olfactory bulb.

During the wine-tasting process, it is important to breathe air through the wine in order to vaporise the volatile compounds. Accordingly, a free passage is needed to and from the nose to enable transmission. If a person has a blocked nose, it is not just the sense of smell that disappears but most of the sense of taste. The taste sensations can be increased if we breathe out via the nose after we have spat out the wine, and this action will help determine the 'finish' of the wine.

As we take wine into the mouth, taste, chew and dissect it before finally spitting, numerous sensations develop. This evolution may be considered in stages: the initial *attack* as the wine is taken into the mouth, followed by the *development* on the palate where we perceive the flavour characteristics and intensity, and then the *finish*, which comprises the final impressions of the wine including the balance. The *length* is a measurement of how long the sensations of the finish and aftertaste last. Tasters sometimes refer to the progressive sensations as 'front-palate', 'mid-palate' and 'back-palate'.

Wine Production and Quality, Second Edition. Keith Grainger and Hazel Tattersall.
© 2016 John Wiley & Sons, Ltd. Published 2016 by John Wiley & Sons, Ltd.

19.1 Sweetness/bitterness/acidity/saltiness/umami

The sensory cells are contained within the 5000 or more taste buds on the tongue – young people may have up to 10,000 active taste buds. There are also some taste buds on the roof of the mouth and back of the throat (which is perhaps why a few people claim to only get 100% of what the wine has to offer if they swallow, not spit). Although highly sensitive, the receptors of the taste buds can only detect five basic tastes: sweetness, bitterness, saltiness, acidity and umami (the savoury taste of amino acids). These are the non-volatile compounds present in wine (although it should be remembered that acetic acid is volatile). There are claims of a sixth basic taste, the bitter, chalky taste of calcium, and of other basic tastes, but such claims remain highly controversial. Of the five basic tastes, saltiness (comprising mainly sodium chloride) is not important in wine. The sensory cells of the tongue convert the detected tastes into electrical signals and send them to the brain's taste cortex. Until fairly recently, it was generally accepted that different parts of the tongue detect these basic tastes. Many wine-tasting books and human biology texts illustrate a diagram of the tongue detailing these areas. However, as we saw in Chapter 16, this concept was discredited by Linda Bartoshuk. For the purposes of this chapter, we will rely on the approach that defined areas of the tongue are more sensitive to the individual basic tastes: it is not disputed that the 'traditional' areas of detection do identify the tastes, only that the other areas do not. It is also accepted that the centre part of the tongue is much less sensitive to the basic tastes. There are tactile sensations of the wine that are also detected in the mouth and on the cheeks, teeth and gums, and these include tannin, body and alcohol.

When assessing the palate of a wine, we consider the following headings:
- dryness/sweetness;
- acidity;
- tannin;
- alcohol;
- body;
- flavour intensity;
- flavour characteristics;
- other observations;
- finish.

19.2 Dryness/sweetness

Before discussing perceptions of sweetness, we need to briefly revisit the topic of grape sugars. Grapes contain glucose (grape sugar) and fructose (fruit sugar), which will be completely or partially converted to ethanol and carbon dioxide during the fermentation process by the action of yeasts. As we saw in Chapter 9, if there is insufficient natural sugar in grapes to produce a wine of the required alcoholic degree, in some countries the winemaker may add sucrose to the must, a process generally known as chaptalisation. In theory, any added sucrose should be fermented to dryness. Although often practised in many of the wine-producing member states of the EU, particularly in years of poor weather, there has long been discussion concerning the possible banning of the process, especially following a proposal adopted by the European Commission in July 2007. However, at the time of writing, the process remains permitted in the more northerly zones of the EU. An alternative to chaptalisation is the addition of concentrated grape must, which of course comprises largely glucose and fructose.

Sweetness, if any, in a wine will be particularly detected on the tip of the tongue. It is important to remember that we cannot smell sweetness (sugar is not volatile), although the nose of some wines may lead us to expect that they will taste sweet. This may or may not be the case – for example, a wine made from one of the family of Muscat varieties may have a fragrant and aromatic nose reminiscent of sweet table grapes, but the wine, when tasted, may be bone dry. Other characteristics can also give an illusion of sweetness, in particular high alcohol levels (although too much alcohol can lead to a bitter taste) and vanillin oak. Thus, a high-alcohol wine that has undergone oak treatment can mislead the taster into perceiving that it is sweeter than the actual level of residual sugar. Pinching the nose while rolling the wine over the tip of the tongue can help the novice overcome any distortions that the nose may be giving. However, the acidity of the wine also impacts on the taster's perception of sweetness – the higher the acidity, the less sweet a wine containing residual sugar may appear to be.

Thresholds for detecting sweetness vary according to the individual: nearly 50% of tasters can detect sugar at a concentration of 1 g/l or less, with just 5% unable to detect sugars at less than 4 g/l.

Residual sugar in a wine is due to fructose remaining after the fermentation. The level of residual sugar in white wine can range from 0.4 to 300 g/l. Most red wines are fermented to dryness or close to dryness, i.e. between 0.2 and 3 g/l of residual sugar. However, because dry wines are very fashionable, some wines are labelled or described as 'dry' when they are anything but. It is common for many New World branded Chardonnays to have between 5 and 10 g/l of sugar, and branded New World reds may contain up to 8 g/l, the sugar helping to soften any bitterness, i.e. a little sweetness in a red wine can serve to balance any phenolic astringency. Wolf Blass, former owner of the famous Australian winery (owned at the time of writing by the somewhat troubled Treasury Wine Estates), is quoted as saying: 'to sell your wine in Great Britain you must do two things: label it dry and make it medium!'

We may consider and note sweetness on the following scale:

dry – off-dry – medium-dry – medium-sweet – sweet – luscious

19.3 Acidity

Acidity is particularly detected on the sides of the tongues and cheeks as a sharp, lively, tingling sensation. Medium and high levels of acidity encourage the mouth to salivate.

All wines contain acidity: whites generally more than reds, and those from cooler climates more than those from hotter regions. In the ripening process, as sugar levels increase, acidity levels fall – mostly a reduction in malic acid, and pH increases. Thus, a cool-climate white wine might have a pH of 2.8, while in a hot climate, a red wine might have a pH as high as 4. Uniquely among fruits of European origin, grapes contain tartaric acid, and this is the main wine acid, although malic and citric are also important. These three acids account for over 90% of the level of acidity. Other acids present may include lactic, ascorbic, sorbic, succinic, gluconic and acetic acids. An excess of the volatile acetic acid is most undesirable. At very high levels, it imparts a nose and taste of vinegar. If the grape must has insufficient acidity, the winemaker may be allowed to add acid, usually in the form of tartaric acid. Within the EU, such additions are only permitted in warmer, southern regions.

Perception thresholds for acidity vary according to the individual. Nearly 50% of tasters are able to detect tartaric acid in concentrations of 0.1 g/l or less, and the remainder at between 0.1 and 0.2 g/l. However, sweetness negates the impact of acidity, and vice versa, and the relationship between these is one of the considerations when considering *balance* as discussed below.

A wine's acidity may be assessed and described on the following scale:

low – medium (–) – medium – medium (+) – high

19.4 Tannin

Tannin is mostly detected by tactile sensations, particularly making the teeth and gums feel dry, furry and gritty. The sensations can be mouth puckering, and after tasting wines high in tannin, you want to run your tongue across the teeth to clean them. Hard, unripe tannins also taste bitter and 'green'. Tannin is a key component of the structure of classic red wines and gives 'grip' and solidity.

Tannins are polyphenols, the primary source in wine being the skins of the grapes. Stalks also contain tannins of a greener, harder nature and nowadays are, with some notable exceptions, generally excluded from the winemaking process. Oak is another source of tannin, and wines matured in new or young barrels, or otherwise oaked, will absorb tannin from the wood.

Tea contains tannin, and a good way of tasting the effect of various tannin levels without the influence of possible distorting factors such as acidity and alcohol is to make several cups of tea (no milk or sugar) with different amounts of maceration as shown below:

- 1 tea bag, 30 s maceration;
- 1 tea bag, 1 min maceration;
- 1 tea bag, 2 min maceration;
- 2 tea bags, 2 min maceration;
- 3 tea bags, 2 min maceration.

Allow the tea to cool and then taste, ensuring that the liquid is thrown onto the teeth and gums. Note the increase in the gritty, drying sensations on the teeth and gums with each taste of stronger tea.

Tannin binds and precipitates protein. This, of course, is one of the reasons why, in general, red wines match red meats and cheeses successfully. This combination causes wines containing tannin to congeal into strings or chains as it combines with protein in the mouth, and thus our perception of tannin in a wine will change if we keep it in the mouth too long. To observe this, take a good mouthful of red wine low in tannin such as a Beaujolais, chew it for 20 s or so and then spit out into a white bowl. Now, repeat the exercise with a red wine high in tannin such as a Barolo. It will be observed that the greater the tannin level in the wine, the more the wine will have formed these strings. Novice tasters often confuse the sensations of acidity and tannin. A classic Barolo, which is high in both, is a good example to taste to distinguish between them. The tannin gives the dry, astringent sensations on the teeth, gums and even hard palate. The acidity produces the tingling sensations on the sides of the tongue and cheeks.

It is often written that white wines contain no tannin. This is not true, although generally the levels are low compared with red wines. The grapes for white wines are pressed pre-fermentation, the solids are settled or the must otherwise clarified, and reasonably clear juice is fermented. Unless there is any period of skin contact, post-crusher and pre-press, the phenolics in the skins will have limited impact. In the case of whole cluster pressing, the role of phenolics is minimal. White wines that have been fermented or matured in oak barrels (or otherwise oaked) may contain considerable oak tannins.

The quantity of tannins in white wines ranges from 40 to 1300 mg/l, with an average of 360 mg/l. Red wines contain 190–3900 mg/l, with an average of 2000 mg/l. Thus, it will be seen that while the average tannin level of red wines is six times that of whites, many white wines contain considerably more tannins than some reds. It is legal in most countries, including member states of the EU, for winemakers to add grape tannin, usually in powder form during the fermentation process, and this is often done to give an over-soft red, a little more 'grip'.

Wine tannins should be assessed and noted for both the level and nature. We may assess the level on the following scale:

low – medium (–) – medium – medium (+) – high

The nature of the tannins may be described as *ripe/soft* or conversely *unripe/green/stalky*. Texture, too, comes under this heading – tannins may be described as *coarse* that is rough, assertive and very gritty, or *fine grained* for those with a smooth, velvety texture.

19.5 Alcohol

Alcohol is detected on the palate as a warming sensation on the back of the tongue and the cheeks, and in the mouth generally. The weight of the wine in the mouth also increases with higher alcohol levels. Over-alcoholic wines will even give burning sensations.

The alcoholic content of light, i.e. unfortified, wines ranges from 7.5% to 16% abv (alcohol by volume). As grapes ripen, the levels of fructose and glucose increase, thus increasing the potential amount of alcohol. So, we would expect wines from hotter climates to contain more alcohol than those from cooler regions. However, it is worth considering at this point that the amount of alcohol by volume in a wine, as is the case with all other aspects of style, flavour and quality, will depend on numerous factors:

- the *climate* of the region of origin;
- the *weather* affecting the vintage in question;
- the *grape variety* or varieties used;
- the *soil* type and drainage;
- *viticultural practices* – including yield, training system, canopy management and choice of harvest time;
- *vinification techniques* – including fruit selection, must concentration (if employed), yeast type, fermentation temperatures and whether or not an incomplete fermentation is stopped, or the wine is blended with unfermented grape juice (*süssreserve*).

The average alcohol level of wines has increased during the last two or three decades. This is due to growers delaying harvesting until phenolic ripeness is reached (especially as the market now demands a softer style of reds than in the past), the use of alcohol tolerant yeasts and the impact of climate change. In 1989, referring to Australian Cabernet Sauvignon, Bryce Rankine wrote in the reference work *Making Good Wine*: 'A ripeness of 10°–12° Baumé (18–21.6° Brix) is usual, which results in wine containing between about 10 and 12% alcohol by volume'. In 2015, any Australian Cabernet Sauvignon with less than 13% abv would be regarded as atypical – over-cropped and under-ripe. As we have seen, there are methods of removing alcohol from over-alcoholic wines including reverse osmosis and spinning cones, but these remain controversial.

We may assess the level of alcohol on the following scale:

low – medium (–) – medium – medium (+) – high

If a fortified wine is being assessed, the alcohol will be in the range of 15–22% abv. We assess whether the wine has been fortified to a low level (15–16% abv), e.g. Fino Sherry or Muscat de Beaumes-de-Venise, a medium level (17–19% abv), e.g. Sherries other than Fino/Manazanilla, some *vin doux naturels* or a high level (20% abv or more), e.g. Ports.

19.6 Body

Body, sometimes referred to as weight or mouth feel, is more of a tactile than a taste sensation. It is a loose term to describe the lightness or fullness of the wine in the mouth. Body should not be confused with alcohol, although it is unlikely that a wine low in alcohol will be full bodied. Generally, wines from cooler climates tend to be lighter-bodied than those from hotter areas. However, as with the other aspects of style, flavour and quality, the body of a wine will depend on the factors listed in Section 19.5. Certain grape varieties usually produce light-bodied wines; others full-bodied ones. Although it is a huge generalisation, wines made from Sauvignon Blanc or Riesling tend to be fairly light in body, while those made from Chardonnay or Viognier may be medium to full-bodied. Of red grape varieties, Pinot Noir usually produces a lighter-bodied wine than Cabernet Sauvignon or Syrah.

The body of a wine is supported by its *structure*, which is made up of a combination of acidity, alcohol, tannin (red wines) and any sweetness. The structure may perhaps be thought of as the architecture of a wine.

Body may be assessed on the following scale:

light – medium (–) – medium – medium (+) – full

19.7 Flavour intensity

Flavour intensity should not be confused with body. A wine can be light bodied but with very pronounced intensity of flavour, for example, a fine Riesling from Germany's Mosel. However, as with the other aspects of style and quality, the flavour intensity of a wine will depend on the factors listed in Section 19.5. Of particular significance is the yield in the vineyard. The flavours of wines from high-yielding vines are generally more dilute and lack the concentration of those from vines with a low yield, although a low yield for one grape variety, e.g. Sauvignon Blanc, would be considered high for another, e.g. Pinot Noir. The impact of yield is discussed further in Chapter 25. Flavour intensity is one of the key considerations when assessing quality.

Flavour intensity may be considered on the following scale:

light – medium (–) – medium – medium (+) – pronounced

19.8 Flavour characteristics

As with the nose, we may consider flavour characteristics on the palate in five basic groups:

fruit – floral – spice – vegetal – other

The reader is again referred to Table 18.2. Detailed notes should be made on the individual flavours perceived, and when noting individual terms these may be linked to known varietal characters. Table 19.1 lists some of the flavours that may be perceived in white wines and the grape varieties or other wine components commonly associated with them. Table 19.2 lists some of the flavours found in red wines and their associated varieties and other wine components.

19.9 Other observations

The *texture* of a wine should be considered, and crucially the *balance* of the characteristics already discussed in this section. The easiest way to understand texture is by imagining running the tips of your fingers over the skin of various parts of the body of people of different ages and professions: the smooth, soft face of a model, the hands of a cashier, the weathered face of a deep-sea fisherman, the pre-shave chin of a builder. The texture of a wine might be described in the range:

silky – velvety – smooth – coarse

Bubbles or spritz, if present, give tactile sensations on the tongue. In a poor-quality sparkling wine, they are very aggressive, while a creamy feeling mousse is indicative of the well-integrated carbon dioxide in a good-quality example. Thus, in the case of sparkling wines, the *mousse* might be described on the scale of

delicate – creamy – aggressive

Balance is the interrelationship between all the taste and tactile sensations, and the components that create them. If any one or a small number of them

Table 19.1 Some white wine flavours and the grape varieties or other wine components commonly associated with them

Apple	Chardonnay (cool climate), Riesling
Apricot	Riesling, Viognier
Asparagus	Sauvignon Blanc
Banana	Chardonnay (hot climate)
Butter	Malolactic fermentation completed
'Catty'	Sauvignon Blanc
Citrus	Chardonnay (cool climate), Riesling
Coconut	Oak ageing
Cream	Malolactic fermentation completed
Creamy texture	Lees ageing
Elderflower	Sauvignon Blanc
Gooseberry	Sauvignon Blanc
Grapefruit	Chardonnay, Sémillon
'Herbaceous'	Sauvignon Blanc
Herbs	Pinot Grigio
Honey	Chenin Blanc, Riesling, Viognier
Kerosene	Riesling (aged)
Kiwi	Pinot Grigio, Sauvignon Blanc
Lanolin	Sémillon
Lemon	Chardonnay, Pinot Grigio
Lime	Riesling (moderate climate), Sauvignon Blanc
Lychee	Gewürztraminer
Mandarin	Sémillon
Mango	Chardonnay (hot climate)
Melon	Chardonnay (moderate climate)
Nectarine	Sémillon
Nettles	Sauvignon Blanc
Nuts	Chenin Blanc, Oak ageing
Passion fruit	Sauvignon Blanc
Peach	Chardonnay (moderate climate), Riesling, Chenin Blanc
Pear	Chardonnay (cool climate), Pinot Grigio
Pepper – bell (green)	Sauvignon Blanc
Petrol	Riesling (aged)
Pineapple	Chardonnay (hot climate)
Roses	Gewürztraminer
TCP	Noble rot (*Botrytis cinerea*)
Toast	Oak ageing
Vanilla	Oak ageing (especially American oak)
'Wet wool'	Chenin Blanc

dominate, or if there is a deficiency of any of them, the wine is unbalanced. An easy-to-understand example is that a white wine described as sweet or luscious but with a low acidity will be flabby and cloying – in other words, unbalanced. A red wine with light body, light flavour intensity, medium (–) alcohol but high tannin will feel very hard and astringent, again unbalanced. A balanced wine has all the sensations in proportions that make the wine

Table 19.2 Some red wine flavours and the grape varieties or other wine components commonly associated with them

Animal	Pinot Noir, high level of *Brettanomyces*
Aniseed	Malbec
Banana	Gamay, carbonic maceration
Blackberry	Grenache, Merlot, Shiraz (Syrah)
Blackcurrant	Cabernet Sauvignon
Bramble	Zinfandel
Cedar	Oak ageing, Cabernet Sauvignon
Cherry – black	Cabernet Sauvignon, Merlot, Pinot Noir (very ripe)
Cherry – red	Pinot Noir (fully ripe), Sangiovese, Tempranillo
Chocolate – dark	Cabernet Sauvignon, Shiraz
Cinnamon	Cabernet Sauvignon
Clove	Grenache
Coconut	Oak ageing
Game	Pinot Noir
Grass	Unripe grapes
Herb (mixed)	Grenache, Merlot, Sangiovese
Leafy	Pinot Noir
Leather	Shiraz (Syrah), aged wines
Liquorice	Grenache, Malbec, Cabernet Sauvignon
Meat	Pinotage
Metal	Cabernet Franc
Mint	Cabernet Sauvignon (especially cool climate)
Pencil shavings	Cabernet Sauvignon
Pepper – bell	Cabernet Sauvignon, Carmenère
Pepper – ground black	Shiraz (Syrah)
Pepper – white	Grenache
Plum – black	Merlot
Plum – red	Merlot, Pinot Noir (overripe)
Redcurrant	Barbera
Raspberry	Cabernet Franc, Grenache, Pinot Noir (ripe)
Roses	Nebbiolo
Smoke	Oak ageing
Soy sauce	Carmenère (very ripe)
Stalky	Wet vintage, unripe grapes, inclusion of stalks
Strawberry	Grenache, Pinot Noir (just ripe), Merlot, Tempranillo
Tar	Nebbiolo, Shiraz
Tea	Merlot
Toffee	Merlot
Toast	Oak ageing
Tobacco	Cabernet Sauvignon
Truffle	Nebbiolo

a harmonious whole; in a well-balanced wine, the sensations are seamlessly integrated.

If any component or a number of components are making the wine unbalanced, these should be noted. However, the state of maturity of a wine is an important consideration. While a very-low-quality wine will never be in balance at any stage in its life cycle, a high-quality wine, particularly reds, will often only achieve balance when approaching maturity. Tannins and acidity may dominate in youth, and the taster needs to evaluate all the components and the structure of the wine to anticipate how these will be interrelated at maturity. Undertaking a vertical tasting, including very young and mature examples, of an individual wine from a top-flight producer can help the student understand how and when balance is achieved. Balance is a major consideration when assessing wine quality.

19.10 Finish

Put simply, the length of the finish and aftertaste is the best indicator of wine quality. The terms 'finish', 'aftertaste' and 'length' sometimes give rise to confusion. 'Finish' refers to the final taste sensations of the wine as it is swallowed or spat. 'Aftertaste' encompasses the sensations that remain and develop as we breathe out, while 'length' is the measure of time for which finish and aftertaste last. To determine length, after you have spat out the wine, breathe out slowly, concentrate on the sensations, observing any changes or development, and count the number of seconds that the taste sensations last. The sensations delivered by poor-quality and inexpensive wines will disappear after 5–10 s (short length), and any remaining sensations are likely to be unpleasant. Acceptable-quality wines will have a length of 11–20 s (medium minus to medium length), good wine 20–30 s (medium plus to long length) and outstanding wines a length of 30 s or more (long length). Truly excellent wines may have a length that runs into minutes. It is important that throughout this test of length, the sensations remain in tune with the actual taste of the wine and also that everything remains in balance. The reader may wish to adjust the timing in seconds given above for the various lengths to their own, individual perceptions.

Finish may be considered on the following scale:

short – medium (–) – medium – medium (+) – long

If any unpleasant characteristics dominate the length, obviously they will have a negative effect on perceptions of quality and should be noted accordingly. For example, a wine with unripe tannins and other bitter compounds might have a medium or even long length, but the bitterness will dominate, and the nature of the length becomes increasingly unpleasant.

CHAPTER 20

Tasting conclusions

In this chapter, we consider the conclusions a taster will make, bringing together all of the information gathered during the assessment of appearance, nose and palate. We look at the judgements and value statements a taster may make, the grading of wines and the reasons for, and practicalities of, blind tasting.

20.1 Assessment of quality

20.1.1 Quality level

Quality judgements are framework dependent. This poses a dilemma. Do we consider the quality of a wine within the context of its peer group or against the entire wine world? Can a wine such as a Beaujolais, which is made for early drinking in a soft, immediately approachable style, be described as outstanding quality, even though it is carefully crafted, exquisitely perfumed, expressive of its origin and superior to most others of its type? The key to answering such questions is to be as objective as possible in the assessment and to note the quality of a wine, as perceived by the taster, according to the origin and price levels within which the wine sits.

A wine's quality level may be considered on the following scale:

faulty – poor – acceptable – good – very good – outstanding

20.1.2 Reasons for assessment of quality

It is, of course, very possible that many wines of outstanding quality are not to our palate, and many simple wines may, on occasions, be very appealing. The reasons for our quality judgements should be logical and as objective as

---

Wine Production and Quality, Second Edition. Keith Grainger and Hazel Tattersall.
© 2016 John Wiley & Sons, Ltd. Published 2016 by John Wiley & Sons, Ltd.

possible. When reviewing the tasting assessment, particular consideration should be given to the intensity, concentration and complexity of nose and flavours on the palate, the structure of the wine, the balance and length of finish. The following guidelines should form a framework for quality assessments:

- *faulty* – showing one or more of the faults, detailed in Chapter 21, at a level that makes the wine unpalatable;
- *poor* – a wine that is, and will always be, unbalanced and poorly structured; a wine with light intensity, very simple one-dimensional fruit flavours, maybe some flaws and short finish;
- *acceptable* – straightforward wine with simple fruit, medium (–) or medium intensity, lacking in complexity, medium (–) or medium finish;
- *good* – an absence of faults or flaws, well-balanced, medium (+) or pronounced intensity and with smooth texture, complexity, layers of flavours and development on palate, medium (+) or long finish;
- *very good* – a complex wine, concentrated fruit, very good structure, well balanced, long length of finish;
- *outstanding* – intense fruit, perfect balance, very expressive and complex, classic typicity of its origin and very long length of finish.

A wine of very good or outstanding quality will present the taster with an unbroken 'line', i.e. a continuity from the sensations of attack, when the wine first enters the mouth, through the mid-palate and on to the finish. It will develop and change in the glass and gain complexity; in other words, it will not say all it has to say within a few seconds of the initial nose and taste. An outstanding wine will also exude a clear, definable and individual personality, true to its origin, making a confident statement of time and place. It will excite in a way that seems to go beyond the organoleptic sensations – in other words, it will have the ability to move the taster in a similar way to a work of literature, art or music.

The 'line' of a wine, as detailed above, may be depicted visually in the form of a palate profile. This is a graph that illustrates the intensity and the texture of a wine from the attack (front palate) through to the mid-palate, to the back palate and on to the finish. An example of a palate profile is shown in Figure 20.1.

20.2 Assessment of readiness for drinking/potential for ageing

The topic of when wines are ready for drinking, and the assessment of this during tasting, is, by nature, complex. The life cycle of wines depends on several factors: origin, colour, style, structure and particularly quality. Inexpensive wines are made to be drunk immediately, be they red, rosé or white. The reds

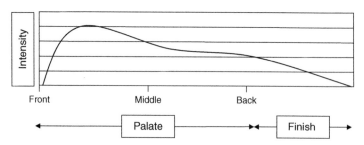

Figure 20.1 An example of a palate profile.

will generally have been made without, or with little, post-fermentation skin maceration that would give a tannic structure, and will have been highly processed, including fine filtration and stabilisation before bottling. Three or four years in the bottle is the maximum keepability, and after this time they will have lost fruit and become 'dried out'. The further we move up the price and quality scales, the more wines benefit from some bottle ageing. Fine red wines are designed for bottle maturation; the time taken for them to reach their peak and just how long they will remain there will vary according to the quality and style of the vintage, the origin of the wine and the winemaking techniques. Intensity of fruit on the nose and flavour on the palate are important indicators, but these may appear closed in youth. The components of solid structure (high tannin, medium to high acidity and appropriate alcohol content) are the keys to a red that will improve in bottle. In youth, these will be unknit, and to the novice, the wine will appear hard and unbalanced, for considerable bottle ageing will be required for them to evolve and integrate. High acidity is a great preservative in particular, but of course, balance is crucial.

20.2.1 Level of readiness for drinking/potential for ageing
All wines have a window of drinkability, beyond which they will be in a state of decline, either slow or rapid, the latter particularly in the case of low-quality wines. Within the window of drinkability, there will, in the case of high-quality wines, be a period when a wine is at its peak, in perfect balance and expressing complex tones that almost defy description. The tertiary characteristics will be fully developed, and the length of finish will be at a maximum. Unfortunately, there can be some disparity in the peak of drinkability between individual bottles of the same wine, particularly if the storage has been inconsistent. As wines decline, the fruit will start to dry out, richness will fade, oxidative and vegetal characteristics may overwhelm, bitterness may come to the fore, and the length of finish may diminish.

It is pertinent and convenient to consider and conclude the level of readiness for drinking/potential for ageing on the following scale:

too young – can drink now but has potential for ageing – drink now: not suitable for ageing or further ageing – too old

20.2.2 Reasons for assessment

Our reasons reaching the above assessment should be included on our tasting note, and we may mention, for example, *structure, balance, concentration, complexity, length* and *typicity.*

20.3 The wine in context

We can now relate the wine to the wider wine world and determine where it sits in terms of its origin, grape variety or varieties, price category and age. The topic of blind tasting is discussed in Section 20.5.

20.3.1 Origins/variety/theme

If the taster has not been informed of the price of the wine being assessed, it may be appropriate to make an estimate. Of course, the conclusion about quality is important, but this has to be related to the origin and maturity of the wine, and an accurate assessment of these is necessary in order to reach a reasoned estimate of price.

20.3.2 Price category

If the taster is aware of the price of the wine being assessed, it is relatively simple to relate the conclusions reached on quality to the price to decide whether the wine is poor, fair or good value for money. However, to some degree, value judgements are also framework dependent, for some wines are, by nature of their origin, scarcity or reputation, more expensive. Of course, at the level of great wines, it is probably not worthwhile attempting to make value judgements. For example, if the wines being assessed are *Prestige Cuvée* Champagnes, costing £120 or more, or *premier cru classé* Bordeaux, which may be priced in many hundreds of pounds, it is pertinent just to make a quality judgement. A poor wine is, of course, poor value, however inexpensive it might be.

We may consider where a wine sits in a range of bands of price on the following scale:

inexpensive – mid-priced – high-priced – premium – super-premium

20.4 Grading wine – the award of points

The grading of wine is a controversial topic. There are those who claim that wines cannot be assessed by scoring points or on a star scale, and that the whole tasting process should be qualitative, not quantitative. Wine crosses the boundaries of art and science, and the exciting and complex characteristics of quality wines cannot be reduced to mere numbers. Countering this argument, many critics point out that in order to show which wines are simply superior to others, they have to be rated on some sort of scale. Critics of musical performances or theatre often give star ratings, as do restaurant inspectors, reviewers of cars, washing machines and pretty much everything that is marketable. However, an opera critic would not dream of rating Puccini's La Traviata 99 points and Tosca 98, or rate an individual performance on such a fine scale. Of course, for the less knowledgeable consumer (or investor), knowing how well a wine has been scored makes the buying decision easier. In recent years, many consumers have become bargain junkies, buying only the latest multibuy or buy-one-get-one-free offer, but others have become points chasers, making their buying decisions on the basis of a points score by a critic.

To confuse matters, there are several different grading systems in use. Historically, a 0–7 scoring scale was often used, and of course star ratings, from 0 to 5 stars, remain a popular system. To many critics, the use of simple scales such as these means that expression cannot be given to the perhaps significant differences in quality between wines in one of the grading bands. Accordingly, the preferred systems now mark the wines out of 20 (with steps that may or may not include fractions) or out of 100. The marks can then, if deemed appropriate, be translated into a star rating. Both systems have strengths and weaknesses, and we will examine them briefly. The scores given can be a sum of the individual scores for appearance, nose, palate, conclusions, etc., and in which case the assessor will mark within a framework or can simply be an overall score. Critics of the former method point out that simply adding the scores means that glaring weaknesses or flaws in one section can mean that a wine is still scored rather highly.

20.4.1 Grading on a 20-point scale

Systems using grading on a 20-point scale are used by judges at many wine shows, and also by several wine magazines, particularly those published outside the Americas. By way of examples, the 20-point system used by the UK-published magazine, *Decanter*, is shown in Table 20.1, and the 20-point system used by the UK-published magazine, *The World of Fine Wine*, is shown in Table 20.2. It will be noted that these two systems are not really compatible.

20.4.2 Grading on a 100-point scale

This is the system used by the world's most influential wine critic, Robert Parker (and his team of tasters), by the US-published *Wine Spectator* and other magazines published in the USA. There are some crucial boundaries, particularly at 90 points, the level at which a wine is perceived as being outstanding. The influence of team Parker on buyers and producers of fine wines, particularly as far as Bordeaux is concerned, cannot be overstated.

Table 20.1 The 20-point scoring system used by the UK-published magazine, *Decanter*

A score of 18.5–20	Outstanding
A score of 17–18.25	Highly recommended
A score of 15–16.75	Recommended
A score of 13–14.75	Fair
A score of 11–12.75	Poor
A score of 10.75 and below	Faulty

Table 20.2 The 20-point scoring system used by the UK-published magazine, *The World of Fine Wine*

A score of 19–20	Great wine of spellbinding beauty and resonance, leaving the drinker with a sense of wonder
A score of 17–18.5	Outstanding wine of great beauty and articulacy
A score of 14.5–16.5	Very good wine with some outstanding features
A score of 12.5–14	Good wine but with no outstanding features
A score of 10.5–12	Enjoyable, simple and straightforward wine
A score of 7.5–10	Sound but dull or boring wine of no character or appeal
A score of 0–7	Disagreeable or faulty wine

It may be said that if Parker gives a wine 89 points, the producer or merchant cannot sell it, and if he gives it 91, the non-millionaire consumer cannot buy it! Using the 100-point system, the lowest score for even the poorest wine is 50, and points are added for appearance (0–5), nose (0–15), palate (0–20), and conclusions, quality and ageing potential (0–10). The system, as used in *Wine Spectator*, is shown in Table 20.3.

If scores are useful at all, it is only to people who understand the system. Much as low scores are obviously not used in marketing material, to the uninitiated a score of 75 out of 100 would equate to 15 out of 20, and both would seem very good! Perhaps there is further confusion if we consider how *The World of Fine Wine* and *Decanter* relate their scores to the 100 point system as shown in Table 20.4.

We will return to this topic in Chapter 22, when we consider the relevance of critical assessments in magazines and annual wine-tasting competitions.

20.5 Blind tasting

20.5.1 Why taste blind?

Tasting wines 'blind' without the taster being given some or all the information about their identity is regarded as the most 'objective' way to assess wine. It is the method used in the tasting competitions discussed in

Table 20.3 Scoring system as used by the USA-published magazine, *Wine Spectator*

A score of 95–100	Classic: a great wine
A score of 90–94	Outstanding: a wine of superior character and style
A score of 85–89	Very good: a wine with special qualities
A score of 80–85	Good: a solid, well-made wine
A score of 75–79	Mediocre, a drinkable wine that may have minor flaws
A score of 50–74	Not recommended

Table 20.4 How *The World of Fine Wine* and *Decanter* relate their scores to the 100-point system

	Points awarded							
Wine Spectator	100	96	90	85	80	70	60	0
Decanter	20	19	17	15.5	14	11	NA	0
The World of Fine Wine	20	19	17	14	11	8	7	0

Chapter 22, as otherwise critics and judges may be swayed by knowing which wine they are tasting, the reputation of the producer and their own previous perceptions of the wine. Blind tasting is also the best way of improving tasting technique, making the tasters rely on their own perception and apply their own descriptors. It is a valuable means to expanding the memory bank, particularly with regard to the relationship between the descriptors and the type of wine tasted. Depending on the objectives, the taster may have absolutely no information or may be given certain relevant details, e.g. the wines are all Burgundy, are all made from one particular variety or are all in a certain price range – in other words, the wines are semi-specified. An alternative approach, sometimes called 'single blind', is when the details of the wines to be tasted are revealed beforehand, but not the order of tasting. Such frameworks can help concentrate the mind in evaluating wines for their quality, typicity, style and maturity.

20.5.2 Blind or sighted?

There is no doubt that knowing the identity of the wine to be tasted impacts on a taster's perceptions of it. There are, however, good reasons why the taster might wish to know the details of the wine being tasted. First, the wine is immediately placed in context, including location within a region, i.e. a sense of place. Second, knowledge of the region and/or producer, e.g. their production methods and 'philosophy', can help the taster understand the wine. Third, knowledge of a wine will help the taster place it at a particular point in its life cycle. Is it at its best? What is the potential to age based on reputation of the producer or region?

20.5.3 Tasting for quality

It can be argued that only by tasting blind can the taster come anywhere near to an objective assessment, as discussed in Introduction to Part 2 – Wine Quality. This is especially important when, considering the factors that reveal quality, including complexity, balance and length, the taster is divorced from being influenced by preconceptions. If the purpose of the tasting is to judge relative qualities, the wines chosen for the event should be comparable from this point of view and, in the broadest sense, be stylistically similar. As we have seen, there is no point in trying to judge the quality of a Beaujolais against a *cru classé* Bordeaux.

20.5.4 Practicalities

When preparing a blind tasting, it is important to ensure that the wines really are presented blind. Unless the glasses are poured out of sight of the participants, the wines should be decanted into numbered neutral bottles. It is not sufficient merely to cover the bottles with foil or bottle sleeves, for the shape and design of the bottle can sometimes reveal nearly as much information as the label. Numbered tasting mats as detailed in Chapter 16 are essential. Depending on the nature of the event, if there are more than eight or 10 wines to be tasted, it is usually sensible to present them in 'flights', as this enables sharper contrasts to be drawn between the individual wines.

20.5.5 Examination tastings

If the tasting is for examination purposes, such as assessing a taster's abilities to accurately describe the wines and deduce the grape varieties, origins, vintages and quality, it is important that the examiner choose representative samples. Each bottle of wine should be tasted by an experienced member of the examination team immediately prior to the examination and detailed notes made. If a wine is discovered to be unexpectedly out of condition, it may be included (provided that any duplicate bottles exhibit the same fault) unless the fault is so severe that it would seriously impact on the other wines tasted.

The taster should be cautious in approaching the wines. The order in which they are presented may not be a sensible order to nose or taste them; for example, a light, delicate dry white will be almost impossible to taste after a rich, sweet one, and a red with a high level of coarsely textured tannins will numb the palate for subsequent wines. While sweetness and tannin cannot be seen or smelt, aspects of the appearance and nose will alert the taster to their probable presence. So, each and every wine should be given an assessment of appearance before proceeding further. A gentle sniff of every wine, too, is essential before the taster decides in which order to undertake a detailed nosing, for a wine with pronounced intensity on the nose may numb the nose somewhat for a subsequent wine with light intensity.

The examination candidate should ensure that tasting notes be made against each of the tasting headings detailed in Chapters 17–19 and in Sections 20.1–20.3. Where appropriate, the notes should be made in detail. Aroma and flavour characteristics in particular require a detailed analysis and description. However, verbosity should be avoided: the examiner is assessing the candidates' ability accurately to describe the wine, not their skills as a wine writer. Each wine should be fully described individually, and not just compared and contrasted with the others. Vague expressions such as 'good', e.g. 'good acidity', and 'reasonable', e.g. 'reasonable tannins', must be avoided.

When attempting to determine grape variety or varieties, origin, quality and maturity, it is necessary to review all of the notes made and ensure that the conclusions are consistent. While first impressions are often correct, a detailed deduction process is necessary in order to consider and review all possibilities. It is important that the tasting notes accurately reflect the wine assessed, and by jumping to a premature conclusion, the candidate may adjust the notes to fit. Students in professional examinations often erroneously believe that reaching the correct conclusion is paramount, while in fact it is upon a detailed and accurate description that the majority of marks are awarded. It is, of course, important that the conclusions be wholly consistent with the taster's descriptions and with each other. They must also be consistent with criteria that describe the wine that the taster has concluded it to be. By way of a very simple example, if the taster has concluded that a wine is a blend of Cabernet Sauvignon and Syrah (Shiraz), it is totally incompatible to conclude also that it is an AC wine from Bordeaux, where Syrah is not a permitted variety. Finally, it is not unknown for students to conclude the identity of the wine erroneously as one that is outside the syllabus or the framework of the examination question!

CHAPTER 21

Wine faults and flaws

Whatever the price point and whatever the quality level aimed for, wines are susceptible to faults. Many of the possible faults appear during the production process and can be prevented or corrected at the appropriate time, but others may manifest themselves only when the wine is in the bottle and sometimes years after bottling. Some faults are so severe as to strip the wine of all semblance of quality or even drinkability.

Wine faults are either chemical or microbiological in nature. However, the issue as to whether an individual wine is faulty (showing one or more serious defects), flawed (showing minor defects) or sound is not necessarily straightforward. Apart from the matter of a taster's individual perception thresholds, there can also be dispute as to whether a particular characteristic is, or is not, a fault or flaw. For example, many Italian red wines have high levels of volatile acidity, which, in wines from other countries, might be considered flaws but which contribute to the very 'Italian' character of the wines. Incidentally, the 1947 vintage of Château Cheval Blanc, Saint-Émilion Premier Grand Cru Classé (A), made before the advent of temperature-controlled fermentations and described by some critics as one of the greatest wines of all time, showed such high levels of volatile acidity that it would today be perceived as faulty. The presence of *Brettanomyces* (Brett) in a wine is another controversial topic: many producers and critics claim it can, at low levels, add complexity, while purists perceive it always to be a fault.

Detailed below are some of the most common faults and flaws encountered in bottled wine. The list is not exhaustive.

Wine Production and Quality, Second Edition. Keith Grainger and Hazel Tattersall.
© 2016 John Wiley & Sons, Ltd. Published 2016 by John Wiley & Sons, Ltd.

21.1 Chloroanisoles and bromoanisoles

Often described as 'cork taint' or 'corkiness', contamination of wine by haloanisoles, in particular chloroanisoles and bromoanisoles, was a relatively common fault, but the incidence has decreased in the last decade. The topic has been, and continues to be, the subject of considerable research. If the wine is heavily contaminated, the fault can be recognised instantly by nosing the wine – there is usually no need to taste. Upon raising the glass to the nose, the smell is immediately one of a damp cellar, a heavy mustiness, a smell of a wet sack, perhaps with tones of mushrooms or dry rot. If the wine is tasted, it will be dirty, fusty and earthy – rather like biting into a rotten apple. The fault is particularly noticeable on the back of the mouth. However, at low levels of contamination, the symptoms detailed above may not be apparent, but the fruitiness of the wine is reduced, and the taste is flat. Thus, contamination by haloanisoles at any level is always a fault.

The main chloroanisoles and bromoanisoles involved in the contamination of wine are as follows:
- 2,4,6-trichloroanisole – perception threshold 1.5–3 ng/l;
- 2,3,4,6-tetrachloroanisole – perception threshold 10–15 ng/l, but as low as 5 ng/l in sparkling wines;
- 2,4,6-tribromoanisole – perception threshold 3.4 ng/l;
- pentachloroanisole – perception threshold >50 µg/l.

Haloanisoles are transformed from the halophenols chlorophenol and bromophenol by the action of micro-organisms, in particular filamentous fungi. One enzyme, chlorophenol-O-methyltransferase, is responsible for the transfer into haloanisoles. Chlorophenols are found in nature owing to anthropogenic sources: they have been widely used as cheap pesticides and fungicides during the past 75 years or so, and as they are not generally degraded by micro-organisms, there is worldwide evidence of their pollutant effects. It is a sobering thought that chloroanisole contamination of wines did not exist before World War II – although Professor George Saintsbury refers to a corked wine in *Notes on a Cellar Book* published in 1920, we cannot know the precise fault noted. Banned in the EU since 1991, chlorophenol usage continues in Africa, Asia and South America. Bromophenols are still permitted in the EU and are used as flame retardants and fungicides, and consequently 2,4,6-tribromophenol has been detected in many production cellars.

In the production of cork wine stoppers, bark may become contaminated with haloanisoles in the forest, either from the atmosphere or from rainwater. Cork may also become contaminated in the production process or later. Thus, historically, all related taints were described as 'corkiness' or 'corked'. However, wines from bottles sealed with closures other than natural cork can also exhibit the fault, thus exposing the myth that contaminated cork is the only cause. 2,4,6-Tribromoanisole taints are derived not from cork but from within the winery. Possible sources of contamination by all the halonanisoles include oak barrels, bungs (especially those made of silicon), pallets, filters, plastics including shrink-wrap as commonly used on pallets of bottles, bottling plants, winery structures and the atmosphere within wineries and cellars. There are several instances of entire cellar buildings having to be demolished and rebuilt owing to contamination. The use of chlorine-based disinfectants and even tap water in the winery can also result in chloroanisole contamination. Haloanisoles are very volatile and migrate easily.

The major manufacturers of corks have undertaken considerable research and made substantial investments in recent years in order to try to eradicate cork contamination. Although this has resulted in a substantial reduction in the problem, cork closures do remain one of the likely reasons for contaminated bottles. Of course, it is impossible for a wine producer to analyse every cork, and a problem may only be discovered once a wine has spent years maturing in the bottle. An interesting point is that cork can act like blotting paper and actually absorb 2,4,6-trichloroanisole from a contaminated wine, thus reducing the level of taint.

There are two other compounds not related to haloanisoles that can give a musty taint to wines: 2-methoxy-3,5-dimethylpyrazine and 2-isopropyl-3-methoxypyrazine.

There has been much debate, at times very heated, about associated taints in recent years, the adversaries being the manufacturers of cork and the manufacturers of alternative bottle closures, and the pro- and anti-cork lobbies. There is divided opinion among trade buyers, writers and journalists. Some supermarkets and other large retailers now insist on synthetic closures or screw-caps when detailing the product specification with their supplier wineries. However, the use of synthetic closures is not always problem-free, as will be discussed later in this chapter.

21.2 Fermentation in the bottle and bacterial spoilage

Fermentation in the bottle may be either alcoholic fermentation or MLF. If a wine contains residual sugar together with live yeast cells, a re-fermentation in the bottle is possible, unless the alcoholic degree is above that at which yeasts work, normally in the region of 15–16% abv. Further, if a wine has not undergone the malolactic fermentation prior to bottling and lactic acid bacteria are present, this may take place in the bottle. The wine will contain bubbles of CO_2, and even if these are not visibly detectable, the CO_2 may be felt as a prickle on the tongue. Active lactic acid can give a haze or silky sheen, particularly apparent when the wine is swirled around the glass. Cloudiness can be indicative of a fermentation in bottle or of bacterial spoilage. A mushroom- or grey-coloured sediment is also a sign of a (completed) bottle fermentation.

21.3 Protein haze

Protein haze is now a very rare fault. The wine will appear dull and oily, as a result of positively charged dissolved proteins being massed into light-dispersing particles. Protein hazes may be removed in the winemaking process by fining with bentonite.

21.4 Oxidation

Oxidation is a fault that is often apparent on appearance and certainly detectable on the nose of a wine. A white wine will look 'flat', not at all bright, and in severe cases will deepen considerably in colour and start to look brown. A red wine will also look dull and take on brown tones. On the nose, the wines will smell burnt and bitter, and have aromas of caramel or, in severe cases, the smell of an Oloroso Sherry. If the wine is tasted, it will be lacking in fruit, bitter, very dirty and short.

At some point in the life of the wine, it has absorbed so much oxygen that its structure has been severely damaged. This may have happened as a result of the grapes having been damaged or having been subject to delays before processing, or owing to careless handling of the wine in the winery, including the failure to keep tanks and barrels topped up, or careless pumping. However, the most common cause is when the wine in bottle has been stored badly or allowed to get too old. Bottles sealed with a natural cork should always be kept lying down to keep the cork moist and expanded in the neck. Wines sealed with plastic 'cork-shaped' stoppers are particularly prone to oxidation after some time in the bottle, as the stoppers harden and shrink. All wines have a finite life and should not be allowed to get too old. As we have seen, inexpensive wines, in particular, are usually made to be drunk almost as soon as they are bottled and are particularly liable to oxidise with extended keeping.

During the late 1990s and early 2000s, there was a high incidence of so-called premox (premature oxidation) in theoretically high-quality and age-worthy white wines, particularly from the Burgundy region. The wines were found to be oxidised and barely drinkable well before their expected maturity. Several theories have been put forward as to the underlying cause, including stressed vines with reduced levels of glutathione (a natural antioxidant), reduced sulfur levels in winemaking and low-quality corks. There are also some indications that premox is affecting some 'blockbuster' red wines, i.e. those with a huge fruit concentration. These may have been made from over-ripe grapes with high pH levels and perhaps when winemakers have given substantial amounts of oxygen, to boost yeast colonies to ensure complete fermentation and soften potential astringency.

21.5 Excessive volatile acidity

The total acidity of a wine is the combination of non-volatile or fixed acids, such as malic and tartaric acids, and the acids that can be separated by steam, the volatile acids. Generally thought of as acetic acid, volatile acid is in fact composed of acetic acid and, to a lesser extent, butyric, formic, lactic and pro-pionic acids. As the name suggests, volatile acidity is the wine acid that can be detected on the nose. All other acids are sensed on the palate. All wines con-tain some volatile acidity. If the level is low, it may increase the complexity of the wine. However, if the level is too high, the wine may smell vinegary or of nail varnish remover (see also Section 21.12). On the palate, the wine will exhibit a loss of fruit and be thin and sharp. The finish will be very harsh and acid, maybe even giving a burning sensation at the back of the mouth.

Acetic acid bacteria, such as those belonging to the genus *Acetobacter*, can multiply very rapidly if winery hygiene is poor, thereby risking an increase in the volatile acidity of contaminated wines. The bacteria can develop in wine at any stage of the winemaking. The bacteria grow in the presence of air. If a red wine is fermented on the skins of the grapes in an open top vat, the cap of grape skins pushed to the top of the vat by the carbon dioxide produced in the fermentation presents an ideal environment for the growth of acetic acid bacteria. The bacteria can also be harboured in poorly cleaned winery equip-ment, especially old wooden barrels. Volatile acidity is also a by-product of the activity of *Brettanomyces* (see Section 21.8). Careful use of sulfur dioxide in the winery inhibits the growth of these bacteria, but overuse amounts to another wine fault as discussed below.

21.6 Excessive sulfur dioxide

Excessive sulfur dioxide can be detected on the nose – a wine may smell of a struck match or burning coke that may drown out much of the fruity nose of the product. A prickly sensation will often be felt at the back of the nose or in the throat. The taster may even be induced to sneeze.

As we have seen, sulfur dioxide is the winemaker's universal antimicrobial agent and antioxidant. As stated above and below, its careful use inhibits the development of acetic acid and *Brettanomyces*. However, excessive use, particularly prior to bottling, can lead to the unpleasant effects described above. It should be noted that the fermentation itself produces some sulfur dioxide. The amount of sulfur dioxide that can be contained in wines sold in the EU member states is strictly regulated by EU wine regulations. Higher levels are permitted for white wines than red, and the permitted level for sweet white wines is greater than that for dry whites. It should be noted that some other countries permit higher levels; for example, the USA and Japan both allow total sulfur dioxide levels of up to 300 mg/l, even in dry red wines, although if sold in the EU, the EU regulations must be adhered to.

21.7 Reductivity

Reductive faults comprise hydrogen sulfide, mercaptans and disulfides. All are recognised on the nose, and all are the result of careless or uninformed winemaking. Hydrogen sulfide has a pronounced smell of rotten eggs or drains. The sensory threshold is in the region of 40 μg/l. Mercaptans can have an even more severe smell, where the odours are those of sweat, rotten cabbage, garlic or even skunk. The sensory threshold is in the region of 1.5 μg/l. Disulfides impart a smell of rubber or even burnt rubber.

Often, sulfur used as a vineyard treatment is reduced to hydrogen sulfide in winemaking by the actions of the yeast. Wines made from grapes that are over-ripe and those grown on poor, nitrogen-deficient soils are prone to reductivity, and some red grape varieties, especially Syrah (Shiraz), are particularly prone. The winemaker should take pains to keep sufficient levels of N_2 and O_2 in the fermenting must to counteract reduction. Diammonium phosphate may be added, usually at a rate of 200 mg/l of must. Its use is common in New World countries, but it does take a little colour out of the wine. Vigorous oxygenation during pump-overs is a proven method to counter the problem. This may involve splashing the wine drawn from the bottom of the fermentation tank into a container, pumping from this to the top of the tank and again splashing it over the cap of grape skins. In the event that a fermented wine exhibits the fault, chemical treatments (e.g. copper sulfate) may be used, but these do impact negatively on the fruitiness of the wine, and careless use can result in a copper haze. Another time-proven remedy is to place a piece of brass into a vat of wine that shows hydrogen sulfide.

Mercaptans are produced after the alcoholic fermentation by yeasts acting upon sulfur or hydrogen sulfide. Racking of red wines immediately after pressing can reduce the risk. In the case of white wines, prolonged lees contact, which can add delightful bread and yeast flavours together with a creamy texture, presents some risk, as the yeasts will scavenge oxygen. Mercaptans can also be suppressed by copper treatments, but disulfides cannot. They are the result of the oxidation and conversion of ethyl mercaptan.

The group includes diethyl disulfide, dimethyl disulfide, dimethyl sulfide and ethyl sulfide.

A topic that has given rise to much research and considerable controversy during the last couple of years is that of post-bottling reductivity, particularly in relation to wines sealed with non-cork closures, especially screw-caps. Numerous incidences of this fault have added more fuel to the ongoing screw-cap-versus-cork debate.

21.8 *Brettanomyces*

Brettanomyces spp. (colloquially known as Brett) are yeasts that resemble *Saccharomyces cerevisiae* but are smaller. They are often present in the skins of grapes and cause a wine defect that is identified on the nose by a pronounced smell of cheese, wet horse, farmyards, baked orange or the plasters used for small wounds. The main compound responsible for imparting the farmyard or manure aroma is 4-ethylphenol. Other compounds are also present, including 4-ethylguaiocol and isobutyric acid, the latter being responsible for imparting cheesy aromas. *Brettanomyces* contamination is often a result of careless winemaking and, particularly, poor hygiene management. Brett is very sensitive to SO_2 and can be inhibited by regular monitoring and additions of SO_2. Barrels can become impregnated with *Brettanomyces* and thus contaminate wine during maturation. Indeed, whole cellars can be contaminated, a situation that is very difficult to rectify. The taste-perception threshold for Brett is approximately 425–600 µg/l. Many producers in the Old World consider that a little Brett can add complexity to a wine and that it is very much an extension of the concept of *terroir*. Pascal Chatonnet, who has undertaken considerable research into contamination by haloanisoles and *Brettanomyces*, states that levels below 400 µg/l can add complexity, but as many as two-thirds of red wines contain levels higher than this. Many wines from the Rhône Valley in France show Brett – historically, a well-documented example was the highly priced and generally very highly regarded Château de Beaucastel, Châteauneuf-du-Pape, but recent vintages have shown no traces. Château Musar from the Lebanon perhaps built its reputation on Brett character. Most New World producers consider any Brett to be a fault and expound that once Brett is allowed to remain in a cellar, the impact on the wines is uncontrollable. That is not to say that there are not numerous examples of New World wines that exude considerable Brett character. Brett is most likely to taint wines that are high in alcohol and low in acidity, with excess nitrogen and with inadequate levels of free sulfur dioxide. There is no universal consensus on the 'correct' level of free sulfur dioxide in red wines, but 25–40 mg/l, depending on the pH, is generally regarded as appropriate. Wines with a little residual sugar are particularly exposed. In recent years the demand for wines with mature, soft tannins has led to many producers waiting for full phenolic ripeness before harvesting. This results in high grape sugars (not always fermented out), high alcohol and high pH levels, which put the wines at risk.

21.9 *Dekkera*

Dekkera is a sporulating form of *Brettanomyces*. It causes defects in wine, recognisable on the nose as a pronounced 'burnt sugar' smell or a 'mousey' off-flavour. It is often a consequence of poor hygiene in winemaking, especially in the barrel cellar. It can be inhibited by careful use of sulfur dioxide.

21.10 Geraniol

Geraniol ($C_{10}H_{18}O$) is recognisable on the nose, as the wine smells like geraniums or lemon grass. It is a by-product of potassium sorbate, which is occasionally used by winemakers as a preservative or inhibitor of fermentation in the bottle should the wine contain any sugar. It is important to use only the required quantity, or the geraniol defect can arise.

21.11 Geosmin

Geosmin ($C_{12}H_{22}O$) is a compound resulting from the metabolism of some moulds and bacteria including cyanobacteria, commonly known as blue-green algae. It is recognisable on the nose as an earthy smell of beetroot or turnip. Faults that can be attributed to geosmin may be caused by the contamination of barrels with the micro-organisms that give rise to it, or their growth on cork. However, geosmin can also be present on grape clusters, and if 2% or more of the clusters are contaminated, the resulting wine can show the defect.

21.12 Ethyl acetate

Ethyl acetate (CH_3-COO-CH_2-CH_3) is recognisable on the nose as a smell of nail-varnish remover or glue (both of which contain the compound). There can also be a smell of pear drops, but the esters present in the weeks after fermentation also exude this aroma, together with that of bananas. Ethyl acetate occurs when ethanol reacts with acetic acid to produce the compound and water. The usual cause is a prolonged exposure of the wine to oxygen or the inadequate use of sulfur dioxide during the winemaking process. It is present in low concentrations in all wines and can be beneficial to aromas, especially in sweet wines. However, any excess, that is a concentration above the sensory threshold, is considered a major fault. The sensory threshold is approximately 200 mg/l, but this varies with the style of wine.

21.13 Excessive acetaldehyde

Acetaldehyde, formed by the oxidation of ethanol, is present in all wines in small amounts. In high amounts, it imparts aromas similar to some deliberately oxidised wines such as Vin Jaune, and when in excess it gives very burnt aromas.

21.14 *Candida* acetaldehyde

This is a rare flaw or fault in bottled wine. The wine not only smells of straw with tones of fino-Sherry but also exudes a dirty character. A rogue yeast, *Candida vini*, is responsible for the defect. It is a film yeast that can form on the surface of wine in aerobic conditions, and the root cause of the problem is usually a lack of care in keeping vats and barrels topped up.

21.15 Smoke taint

This is a problem found very occasionally in wines from Australia and South Africa as a consequence of bushfires or controlled burning taking place near vineyards. Affected wines may smell of burnt ash, smoked salmon or ashtrays. High concentrations of the compounds guaiacol and 4-methyl-guaiacol are found in tainted wines, but it should be pointed out that wines may contain low concentrations of these as a result of oak maturation or treatments. The sensory threshold is approximately 6 µg/l, but bushfire-affected wine samples submitted to the Australian Wine Research Institute have shown guaiacol levels in excess of 70 µg/l. Of course, growers are aware if their grapes have been affected by smoke, but the fruit is usually harvested even if the wine made is considerably devalued. In Australia, we have been shown, and have tasted from, vats containing 50,000 litres of smoke-tainted wine. When asked about the destination for the product, the winemaker suggested that a certain brand might benefit from the addition of some smoky, toasty flavours in the blend.

CHAPTER 22

Quality – assurances and guarantees

The quality of a wine may be assessed by undertaking a structured and detailed personal tasting, and we can have trust in our own judgements. In this chapter, we discuss whether there are there third-party assurances or guarantees of wine quality upon which we can rely. We consider compliance or otherwise with Protected Designation of Origin (PDO) and Protected Geographical Indication (PGI) legislation, success in tasting competitions, inclusion in official classifications, ISO certification, the quality offered by successful brands and whether high prices guarantee quality.

22.1 Compliance with PDO and PGI legislation as an assurance of quality?

22.1.1 The EU and third countries

Wherever wines are made, the product will to a greater or lesser extent be subject to compliance with local laws and regulations. In some countries, this is largely a matter of ensuring that the product is safe to drink and correctly described on the label. In other countries, and particularly the member states of the EU, the relevant laws are more detailed and more restrictive. All wines produced within member states of the EU must comply with EU wine regulations regarding production, oenological practices including permitted additives and labelling.

The International Organisation of Vine and Wine (OIV) details conditions and limitations of usage of winemaking techniques in the International Code of Oenological Practices. The International Oenological Codex (CODEX) gathers descriptions of the main chemical, organic and gas products used in

Wine Production and Quality, Second Edition. Keith Grainger and Hazel Tattersall.
© 2016 John Wiley & Sons, Ltd. Published 2016 by John Wiley & Sons, Ltd.

the making and keeping of wines. For wines produced or marketed in the EU, the EU Regulations detail permitted practices – these are now in line with CODEX. The current regulations are (EU) No. 479/2008, (EU) No. 606/2009 and (EU) No. 607/2009 which provides details of permitted oenological practices. (EU) No. 607/2009 was subsequently amended by (EU) No. 538/2011 and (EU) No. 670/2011. The production of organic wine is covered in Regulation (EU) No. 203/2012 amending Regulation (EC) No. 889/2008, which lays down detailed rules for the implementation of Council Regulation (EC) No 834/2007, as regards detailed rules on organic wine!

It should be remembered that under EU regulations, everything is prohibited unless specifically permitted. For example, it was only subsequent to Regulation 1507/2006 that pieces of oak wood became a permitted additive to 'quality' wines produced in the EU, a practice long since legally practised elsewhere (and illegally practised within the EU). *Third-country* wines, i.e. those produced outside the EU, must comply with the regulations if they are to be exported to the EU. Each wine-producing member state of the EU has its own wine laws, which are subordinate to, and comply with, the requirements of the EU regulations.

22.1.2 PDO, PGI and wine
Wines produced in the member states of the EU are divided into three broad categories:
- wine;
- PGI;
- PDO.

Before 2009, the EU categorised wine into:
- table wine without a geographical indication;
- table wine with a geographical indication;
- quality wines produced in specific regions (QWpsr).

Some of the names that individual countries gave to the table wine with a geographical indication category were well known: most drinkers were aware of *Vin de Pays* (France), but few had heard of *Deutscher Landwein* (Germany). The names given by countries to individual PGI wines are similar to those previously given to the previous table wine with a geographical indication, e.g. Vin de Pays des Côtes de Gascogne has simply become Côtes de Gascogne IGP.

22.1.2.1 PGI wines
According to EU Regulations, PGI means an indication referring to a region, a specific place or, in exceptional cases, a country, used to describe a wine that complies with the following requirements:
1 it possesses a specific quality, reputation or other characteristics attributable to that geographical origin;

2 at least 85% of the grapes used for its production come exclusively from this geographical area;

3 its production takes place in this geographical area;

4 it is obtained from vine varieties belonging to *Vitis vinifera or* a cross between the *Vitis vinifera* species and other species of the genus *Vitis*.

The authors of this book note that this last point means that hybrid varieties are permitted for PGI wines. These are *vinifera* crossed with an American species, and are not usually noted for their taste attributes! It is also noted that Article 25 of (EU) 607/2009 provides that PGI wines are subject to an analytic test and *may* have an organoleptic test. In other words, a PGI wine must be technically sound but could taste dreadful! Having said that, most PGI wines are of acceptable quality and, in a number of cases, are better than some PDOs.

22.1.2.2 PDO wines

The concept of PDO is simple, although the detail contained in the regulations is not. Wines produced in a region, or more tightly defined area, should be typical and distinctive of the defined origin. The consumer does not want or expect a red Bordeaux to taste like a red Burgundy, and we would expect a white wine from Mosel to have different characteristics to one from Baden. Each wine-producing member state that has an EU-recognised PDO wine regime has its own name(s) for the PDO category or categories, together with detailed rules of production and labelling, always within the already-precise rules of the EU regulations. The question is, do these rules guarantee that a wine designated and labelled as a quality wine is just that? Indeed, in the case of Italy which (as with Germany and Spain) divides its PDOs into two broad categories, the 'highest' level is called 'Denominazione di Origine Controllata e Garantita' (DOCG), certainly implying guarantee of quality.

Under the EU regulations, a 'designation of origin' means the name of a region, a specific place or, in exceptional cases, a country used to describe a product referred to in Article 33(1) that complies with the following requirements:

1 its quality and characteristics are essentially or exclusively due to a particular geographical environment with its inherent natural and human factors;

2 the grapes from which it is produced come exclusively from this geographical area;

3 its production takes place in this geographical area;

4 it is obtained from vine varieties belonging to *Vitis vinifera*.

Without doubt, it is the French term 'Appellation Contrôlée' (AC), now also known as Appellation d'Origine Protégée (AOP), that is the best-known PDO to wine lovers and, with apologies to other member states of the EU, we will limit our discussion here to France.

22.1.3 The concept of AOP (AC)

AOP (AC) is, in theory at least, a guarantee of origin and a very basic typicality. It should be noted that AOP applies not only to wines but also to many French foods, e.g. AOP Roquefort (cheese). The system is administered by the Institut National de l'Origine et de la Qualité. Factors controlled by AOP regulations include the following:

- delimitation of vineyard area;
- permitted grape varieties;
- maximum yield per hectare;
- methods of viticulture;
- methods of vinification;
- minimum alcoholic degree of the wine;
- wine must pass a tasting test;
- wine must pass a laboratory analysis.

Essential to the concept of AOP is the delimitation of the area of production. The area can be any of the following:

- a *region*, e.g. Bourgogne AOP (Burgundy);
- a *district* within a region, e.g. Chablis AOP;
- a *group of villages*, e.g. Côte de Beaune Villages AOP;
- a *commune* (Parish), e.g. Beaune AOP;
- an *individual 'premier cru' vineyard*, e.g. Beaune Toussaints Premier Cru AOP;
- an *individual 'grand cru' vineyard*, e.g. Corton Charlemagne Grand Cru AOP.

In theory, the more precise the appellation, the more individual the wine should be. It is where the grapes are grown that determines the AOP, not where the wine is made, e.g. wine made from grapes grown in Chablis will bear the AOP Chablis, even if the wine was made elsewhere in Burgundy, e.g. the Côte d'Or (although the wine must be made within the region).

As stated, AOP is not a guarantee of quality, but the more precise the Appellation, the better the land is officially rated for its potential for higher-quality production and the more strict are the legal criteria that apply, which *may* result in better quality. Much depends on the producers and their commitment to quality. One of the drawbacks of the system is that individual appellations have a commodity value. For example, in Chablis, a wine from an individual Grand Cru vineyard usually has a value of double or treble a wine that just bears the district appellation, Chablis AOP, however good the latter might be. This of course does little to encourage producers of the 'lower' appellations to make the very best wine possible from their land.

As we have seen, it is not only the vineyard area that is delimited for AOP wines. There are many other factors that will contribute to taste, style and individuality, and these are embodied in the AOP laws. One of the key factors is, of course, the grape variety or grape varieties from which the wine is made. In some regions or districts, the AOP laws restrict the possibilities to just one variety (e.g. all Chablis AOP and nearly all other white Burgundy must be made from Chardonnay), while in other areas, the grower has a wider choice and blends the wine from a number of different varieties. However, one of the key points is that generally the varieties permitted for AOP wines are usually those that are traditional to the area. If growers wish to use a non-traditional variety, they will generally not be able to sell the wine with the name of an AOP although an IGP may be possible, and the Vin de France category will be available to them.

The density of vine planting may be specified in the AOP regulations. For example, in the Bordeaux region, vines planted in the communal Pauillac or Saint-Estèphe' appellations must be planted at a minimum density of 6500 vines per hectare. For the district appellations of Médoc or Graves, a minimum of 5000 vines per hectare is required, and under the latest changes, for the regional Bordeaux Supérieur or Bordeaux ACs, 4500 and 4000 vines per hectare, respectively.

Article 25 of (EU) 607/2009 provides that PDO wines are subject to an analytical and organoleptic test. The tasting and laboratory analytical tests for Appellation Contrôlée wines were historically, in many regions including Bordeaux, not of the finished wines. The wines were assessed part way through their development. In other words, the wines tasted were never marketed as such, for after the tasting, they would have been subject to further ageing, in vat or barrel, blending and preparation for bottling, including perhaps fining, filtering and cold stabilisation. It was argued that such a tasting test did little to guarantee that the purchaser would receive even a typical and acceptable wine, let alone a quality product. However, from 2008, changes in the legislation required the wines to be assessed at bottling time. The vast majority of wines are accepted, and it is rare that a wine is declassified or rejected. There is, of course, an appeals process for those that are. However, the assessment committees do reject some wines of high quality. Les Hauts de Pontet-Canet, the 'second' wine of the 2012 vintage of the top-performing Bordeaux cru classé Château Pontet-Canet, was rejected for the AC (AOP) Pauillac by the authorities when assessed in 2014. The legal status was reduced to Vin de France, officially classified by the EU just as 'wine'. This wine, which has been tasted by the present authors, is indeed excellent, and such demotion is possibly on account of the tasting panel not accepting it as typical of the appellation Pauillac. Other notable producers who have had wines rejected include the famed Muscadet producer Domaine de l'écu. There are now many high-quality producers who choose to boycott the AOP or equivalent PDO regulations. Their reasons are numerous: for example, they may plant or even retain the 'wrong' grape variety, for the varietal requirements do sometimes change. Or they may choose not to comply with other aspects of the legislation, such as density of planting or cultural techniques. The resulting wines can indeed be excellent, if sometimes atypical. Without doubt, there are many poor wines bearing an AOP (AC) on the label, and many excellent wines simply sold as 'Vin de France' (or other national equivalent).

It is worth noting here that producers in third countries sometimes borrow the names and labelling terms of perceived illustrious appellations, although such labelled wines may not be imported into the EU. Fortunately, after pressure from the EU, the use in Australia and USA of terms such as 'Burgundy' and 'Chablis' ceased some years ago, but 'port' lingers, especially in California and South Africa. The terms 'premier cru' and 'grand cru' legally protected and with particular significance in France mean nothing in the New World: in South Africa, there are several wines marketed under the label 'premier grand cru'. At the time of writing, they sell for around ZAR 36 (approximately £1.90 at the then rate of exchange) per bottle – one is described in the *John Platter South African Wine Guide* as 'mildly fruity'.

22.2 Tasting competitions and critical scores as an assessment of quality?

The assessment of the quality of various wines by a columnist for the purpose of conveying the results and giving the consumer a 'best buy list' is a regular feature of weekly newspaper columns, monthly magazine articles and annual wine guides. Some of these opinion formers are expert tasters and well educated in vinous matters, others sadly are not. As we saw in Chapter 20, there are also tastings undertaken by panels of expert tasters for wine magazines and annual tasting competitions such as the *International Wine Challenge, Decanter World Wine Awards* and the *International Wine and Spirit Competition*. The results are categorised by point tables or the award of medals. Such tastings are usually carried out under blind conditions (although this is not the case for the tastings for Robert Parker's *Wine Advocate* or the annual *John Platter South African Wine Guide*) and can be a useful source of information for the prospective purchaser. However, there are several caveats, particularly with regard to annual wine competitions:

1 Especially at the large competitions, where tasting panellists assess large numbers of wines, it can be the biggest, loudest wines that win through. It is sometimes said of Australian wine that the more medals that appear on the bottle, the less likely you are to want to drink the whole bottle.
2 The wines tasted will have been subject to recent movement, and some might not be shown in the best of condition.
3 Annual wine competitions require entry fees from the producers/agents, and for this and many other reasons, many decide not to enter. There is little to be gained for a producer entering a wine that already has a high reputation.

4 In some competitions, the majority of wines win medals. The more sceptical reader might think that this is on account of the completion organisers looking for payments from the medal winners for bottle stickers and other promotional material. However, Italy's Concorso Enologico Internazionale is very sparing with its medals, awarding these to approximately 5% of the wines entered.

5 Perhaps most importantly, the assessment is merely a snapshot of the particular bottles tasted at that time. Large producers may keep many of their wines, especially whites and rosés, in vat until such time as they are needed for the market. Bottling is expensive, and bottled stock is space-consuming. Further, many wines may be kept 'fresher' in vat – this is especially true of aromatic white varieties such as Sauvignon Blanc. Accordingly, there will be several bottlings taking place of 'medal-winning' wines throughout the year. At very least, each wine will be adjusted and stabilised for bottling, but vats may have developed differently and blending may take place before bottling. So, the wine being shipped may not be identical to the medal-winning wine.

Finally, in this section, it is worth noting that critics and judges may have strong views on how wines should taste, and there can be considerable dissension. In 2014, a panel of three expert tasters (including two Masters of Wine) assessed a range of Oloroso style wines for The World of Fine Wine Magazine. The individual panellists scored an Osborne Oloroso Solera India 19, 17 and 15 (out of 20), and a Delgado Oloroso 14.5, 13 and 5, respectively. In another panel tasting of Jura wines for a different issue of the same magazine, the individual scores for 2011 Maison Pierre Overnoy/Houillon Arbois Pupillin were 15.5, 8 and 3.5. The dissenting views of two influential wine critics on one particular Bordeaux wine, 2003 Château Pavie, Saint-Émilion Premier Grand Cru Classé, are fascinating. Jancis Robinson MW described the wine as 'ridiculous wine, more reminiscent of a late-harvest Zinfandel than a red Bordeaux' and scored it 12/20. Robert Parker described it as 'a wine of sublime richness, minerality, delineation and nobleness' and scored it 98/100. Leading into the next section of this chapter, it is interesting to note that in the 2012 Classification of the wines of Saint-Émilion (discussed below), Château Pavie was one of two properties elevated to Premier Grand Cru Classé (A).

22.3 Classifications as an official assessment of quality?

Classifications, such as the Saint-Émilion classification mentioned above, provide an official assessment of the quality of a property's wine. They are most important in the Bordeaux region. The most famous classification is that of 1855, which classified the red wines of the Médoc district (together with

Château Haut-Brion situated in the district of Graves) and the sweet white wines of Sauternes. This classification was drawn up by wine brokers, based on the prices that the wines achieved and, with the exception of one well-known promotion (Château Mouton-Rothschild in 1973) and one not at all well-known addition to the list (Château Cantemerle in 1861), has not been amended. Several other districts in Bordeaux have since been subject to classification, using many assessment criteria as a basis, including tasting of the wines. The wines of the Graves district were last classified in 1959, and those of Saint-Émilion (which in theory re-classifies every 10 years) in 2012. The Saint-Émilion Classification places properties into two groups: Premier Grand Cru Classés, which is further divided into (A) and (B), and Grand Cru Classés. It should be remembered that the majority of properties remain unclassified.

The recent history and saga of the Saint-Émilion Classification are worth exploring here. The importance to properties of inclusion at the highest possible level of classification is huge. The impact upon the price that can be charged for the wines, and the reputation, image and prestige of the property, cannot be overstated. The value of a property classified as Premier Grand Cru Classé (B) may be 10 times that of an unclassified property. The wines of Saint-Émilion had been classified in 2006, and the next expected classification would have been in 2016. However, the 2006 classification was suspended by a court in March 2007 following complaints from four producers who had been demoted and disputed the result. The classification was subsequently reinstated in November 2007 by France's highest administrative court, the Conseil d'Etat, but incredibly declared invalid by a Bordeaux court on 1 July 2008. Emergency powers to temporally reinstate the 1996 Classification were used by the French government on 11 July 2008, and in May 2009 a law reinstated this classification but allowed the incorporation of properties who had been promoted in 2006. A new classification, using revised criteria, was made in 2012, and, in the short term at least, the dust seems to have settled.

The Cru Bourgeois classification of wines from Bordeaux's Médoc and Haut-Médoc has attracted considerable respect since it was reintroduced in 2008. Each year, there is an assessment by independent panels of wines submitted by the châteaux, and a property that is included one year in the classification list may not be included in other years. For the 2013 vintage, Cru Bourgeois status was awarded to 251 châteaux.

With the possible exception of Cru Bourgeois, the relevance and validity of all the classifications are always subject to debate, especially as owners come and go. Land may be traded within the boundaries of the Appellation, and quality may rise or fall. Interestingly, the superiority of the 1855 First Growths – Château Lafite-Rothschild, Château Latour, Château Margaux, Château Haut-Brion and Château Mouton-Rothschild (added in 1973) (red wine classification) and Château d'Yquem (sweet white wine classification) – is today not really disputed. High prices mean more money for investing in quality, which means higher prices. Those who question the relevance of classifications also point out that the two properties in Bordeaux that now normally achieve the very highest prices, Le Pin and Château Pétrus, lie in the district of Pomerol, which has never been classified. The rise in reputation of Le Pin is staggering – the first year of production under the name was 1979, the year that the property was purchased by the Thienpont family. In 2015, the 1982 vintage commanded price in the region of £4800 per bottle – in 2008, this was a modest £2000 per bottle.

22.4 ISO 9001 certification as an assurance of quality?

ISO 9000 is an identifying code for a series of quality standards published by the International Organisation for Standardisation (ISO), based in Geneva. Central to the series is the standard, ISO 9001:2008. This is a quality system standard that serves as a model for the development, implementation and maintenance of quality management systems. At the time of writing, ISO 9001:2008 is expected to be updated into ISO 9001/2015, which may have fewer prescriptive requirements.

The ISO 9000 series is concerned primarily with what an organisation does to fulfil:

- the customer's quality requirements, and
- applicable regulatory requirements, while aiming to
- enhance customer satisfaction, and
- achieve continual improvement of its performance in pursuit of these objectives.

Those who embrace quality-management systems developed against ISO 9001:2008 state that they may be used by wine producers in the effective control of the production and commercial processes that operate to satisfy customers' needs. Effectively, the quality-management system defines the way in which customers' requirements will be fulfilled and quality improvement achieved. ISO 9001:2008 requires the implementation of a documented quality-management system based on a quality manual, management and operating procedures, and quality records. The use of third-party auditors, or certification bodies, enables the objective and independent assessment of compliance with the international standard and, hence, worldwide recognition of capability and competence in quality management.

ISO 9001 has received criticism from some quarters because of the apparent focus on the achievement of customer satisfaction through the prevention of nonconformity at all production stages. The concept exists that with ISO 9001, rules must not be broken, even if by doing so a superior product might be achieved. Critics believe that with ISO 9001, quality is seen as something that fulfils the customers' requirements, not their aspirations. Breakfast in Ibis hotels in France carries ISO 9001 accreditation! It is thought that quality is dumbed down and, consequently, many top-quality producers reject ISO 9001 and the concepts it stands for. Wine is an agricultural product, subject to the fluctuations of the annual growing season and cannot be consistently excellent. While it is considered that using ISO 9001 is not a guarantee of the quality of the finished product, there are certified producers that unquestionably rely on the standard and believe that by doing so, they produce 'quality' products, however ill-defined the word might be. While visiting a large bodega (winery) in Argentina, the authors asked the chief winemaker about steps they might take to lift the quality of the product. The reply was somewhat frightening: 'our customers can be assured our wines are among the best. We have ISO 9001 and HACCP!' Dewalt Heyns, winemaker at the flagship Saronsberg winery in Tulbagh, South Africa, says: 'the day ISO 9001 happens is the day I retire. You don't get in the vineyard or the winery, but spend your life managing paper. The only benefit is in building the guarantee of tracking and traceability.'

Wineries may be certified as ISO22000:2005, which is an international standard for food-safety management systems. Large customers, such as supermarkets, may request such producer accreditation, as it helps satisfy the requirements for 'due diligence' in matters of food safety. ISO22000 is due to be updated in 2017.

22.5 Established brands as a guarantee of quality?

Consumers' faith in large brands is built around their perceptions that the brands they choose give them a guarantee of quality and consistency, and that the product or service is one that addresses their individual wants and needs. With regard to wine, the successful brands have a large market share and a seemingly ubiquitous presence on the supermarket shelves. However, it is rare to find a wine writer with a kind word to say about the quality and individuality that some of these offer, particularly the super-brands. Many brands have multi-tiered ranges of product pitched at different price levels. An example is *Penfolds* (owned by Treasury Wine Estates), which markets wines that sell from as little as £5 a bottle and up to £350 a bottle or more (Penfolds Grange). It should be recognised that many companies that sell branded wines produce them in more than one country, under the same brand name.

It is not easy to establish the point at which a wine from a long-established producer makes the transition to branded product in the modern sense of the term. For instance, is Château Margaux a brand? Certainly, the well-known

Champagne houses are now established as brands, and the perceptions of quality associated with the Champagne marques are high. But Champagne itself is a relatively small production region (presently 33,705 ha, although this may increase), and the price commanded for the wine means that high standards in the vineyards and cellars are the norm. Grape prices are high, too: in the 2014 harvest, the price paid by the houses was €5.17–6.06/kg (€5170–6060/tonne or £4048–4745 at the exchange rate at that time). The reader may wish to compare this with the prices achieved in the 2013 harvest in South Africa and Australia, as detailed in Chapter 24.

It is interesting that the rise and fall of brands depend as much as if not more on marketing budgets and brand-building activities than any variation in the quality of the product. The biggest-selling brand of South African wine in the UK is *Kumala*. At the time of writing, the brand owner is Accolade Wines, who bought it from Constellation Brands, who had purchased Vincor, who in turn had purchased Western Wines, the original brand owner! It is pertinent to note that until 2007, *Kumala* owned neither a vineyard nor a winery, and the wines were all sourced by one broker. For some time after the acquisition by Constellation Brands, the development of *Kumala* seriously faltered, resulting in a sales decline in the UK from over 3 million cases per year to fewer than 2 million. Mostly as a result of this, sales of South African wine in the UK declined by 11%. In any event, brands, almost without exception, have a finite life. Those drinkers who in the 1970s enjoyed(?) drinking Crown of Crowns, Hirondelle and Don Cortez have long since moved on.

In the last 20 years, New World wine brands have had far more impact in the UK market than those of the Old World. At the time of writing, nine of the top 10 brands for UK off-sales are New World wines. The undisputed largest fine wine region in the world, Bordeaux, with 113 000 ha of vineyard, more than the whole of South Africa, and an average annual production of approximately 800 million bottles, has only one brand (Calvet) in the top 50. To many consumers, the New World generally provides consistent, easy-to-understand wines, very much fruit driven and designed to be instantly approachable.

So, do brands offer the guarantee of consistency and quality referred to in the first paragraph of this section? In order to invest in brand building and maintenance, the owner needs to be reasonably sure of the continuity of the supply of grapes. The concept of blending often has negative connotations among serious wine lovers, even though nearly all wines are blended to some extent. However, blending lies at the very heart of creating and maintaining the style of wines for the big brands. Fruit may be sourced from the producer's own vineyards, growers under contract or the spot market. Wines may be made in one or several wineries, and blenders have many components to maintain the style. However, blending large quantities for consistency, by definition, means that blending cannot be for the best possible quality, for some blends will not meet the hard-to-reach standards.

Blending for consistency means that the individuality of a particular vintage, region or even winemaker (for they come and go) has to be negated. Wine is an agricultural product, and the style and quality of fruit from any region or district will vary according to the weather patterns of the year. Individual properties may decide not to make their 'top' wines in poor years, and the brands that usually source fruit from those districts will face particular challenges. Of course, by striving to be consistent, there may need to be a good deal of technical adjustment to the wines – additions of acid, tannin, etc. Non-branded wines may well have such adjustments, with the purpose of making the particular wine as good as the winemaker can achieve, rather than to maintain the consistency the brand owners require.

Finally, to counter the 'strength of brands is their consistency' argument, the consumer is not guaranteed a consistent product anyway. Tasting conducted by the authors has revealed a considerable diversity in both style and quality of supposedly identical bottles of branded wines. Supermarkets in particular often buy branded product from more than one source. Brands may vary from lot to lot and particularly from market to market. Some of the best-known brands of Champagne produce to a different specification according to the market to which the product is destined. Champagne buyers in Britain generally prefer their Champagnes with much more autolytic ageing on the lees (the legal minimum is 15 months) and with a lower dosage than that required by the buyer in Germany. So, if the supermarket bypasses the regular source of supply and buys via alternative sources, it is possible to find very different cuvées standing side by side on the shelves.

22.6 Price as an indication of quality?

It can be argued that the market is the final arbiter of wine quality. If the product is superior, it will command a higher price than lesser wines. Even before they were bottled in 2007, the finest premier cru classé Bordeaux wines of the excellent 2005 vintage were being traded in London at a bottle price of £450–600, a 50% increase on the price they had been released at less than a year earlier. In the subsequent years, prices rose and fell, owing to many factors including how the wines and properties were seen to be performing and the worldwide financial turmoil. By 2015, the 2005 Château Cheval Blanc was trading at £300–400, well below its release price.

Markets are subjects of hype and fashion. It is interesting to note that in the late nineteenth century, it was the wines of Germany that commanded the highest prices, more than any cru classé Bordeaux. For example, the 1896 wine list of the highly reputed London merchant Berry Bros and Rudd offered:
- 1862 Rüdesheimer Hinterhaus: 200 shillings (£10) per dozen;
- 1870 Château Lafite: 144 shillings (£7.20) per dozen.

The Lafite was the most expensive Bordeaux on offer and from the great 1870 vintage! Even as recently as 1963, Peter Dominic were offering:

- 1959 Château Gruaud Larose: 18 shillings/6d (95 pence) per bottle;
- 1959 Wehlener Sonnenuhr Spätlese (J. Bergweiler): 32 shillings (£1.60) per bottle.

1959 was an excellent year in both Bordeaux and Germany.

In 2015, Fine+Rare offered:

- 2008 Wehlener Sonnenuhr Spätlese (Joh Jos Prüm): £19 per bottle;
- 2005 Château Gruaud Larose: £64.00 per bottle;
- 2005 Château Lafite: £624 per bottle.

It is argued that the price of any wine can be no more than the market will bear. While producers need to recoup production costs and make a profit, this does not mean that the price of the wines necessarily reflects the differences in the production costs of individual products. As we have seen, many brands comprise a range of levels, and often all wines in the level are pitched at the same, or broadly similar, price whatever the production cost. The price will be determined with regard to the perceived market and the image to be maintained. As we have seen, appellations, too, have a 'commodity value', which reflects their image: the producer of the very finest Muscadet can only dream of achieving the price achieved by the maker of a mediocre Sancerre.

CHAPTER 23

The natural factors and a sense of place

In this chapter, we consider the different conceptual styles in which wine may be produced. We look at the concepts of typicity (a word almost unique to the world of fine wine) and regionality. The influence of regional climates, meso- and microclimates and soils and the concept of terroir are discussed. Finally, we consider the 'vintage' factor, the effect of the annual weather and other variables upon the style, quality and quantity of wine made by a producer in any year.

23.1 Conceptual styles

Wines may be considered as being produced in three different conceptual styles:
- variety driven;
- fruit driven;
- site driven.

Grape varieties, to many consumers, are like brands. They offer a statement of style and taste, and by providing a wine that adheres to the benchmark characteristics of the variety of which it is labelled, the expectations are satisfied. As one moves up the quality scale, the wine should be more expressive of the variety. Fruit-driven wines exude masses of primary fruit enhanced, in good wines, by the secondary and tertiary characteristics. Site-driven wines may be more restrained, yet more individual, and express their origins, in all their complexity, rather than a more simplistic statement of variety or fruit.

Wine Production and Quality, Second Edition. Keith Grainger and Hazel Tattersall.
© 2016 John Wiley & Sons, Ltd. Published 2016 by John Wiley & Sons, Ltd.

There are other conceptual styles by which wines may be classified, for example, wines that are:

- expressive of raw materials – grape variety, region;
- made to meet a market – big-volume brands, the creators of which sometimes use consumer panels and focus groups to choose or confirm style;
- winemaker concept wines – often blends of grape varieties or districts – examples include Penfolds Grange, many Bordeaux Cru Classés and 'Super Tuscans'.

23.2 Typicity and regionality

Until 40 years or so ago, the origin of wines was generally regarded as the main factor that determined their style and, to a large extent, their quality. Historically, ripeness or otherwise was accepted as in the hands of the gods, according to the luck of the year. Classic wines such as *cru classé* Bordeaux were fabled for their *goût de terroir* and were made to be laid down for a generation or more – indeed the wines of great but firmly structured vintages such as 1870 and 1928 did not reach maturity for 50 years. Each region's wines were made in the way that they had been for generations. Grape varieties were little discussed – they were much less understood than today, and the varieties grown were those traditional to the area. Chardonnay (historically usually referred to as Pinot Chardonnay) and Pinot Blanc were often confused as the 'white Burgundy noble grape'. By 1968, Viognier was a variety on the verge of extinction, with just 14 ha remaining in France at Condrieu and Château Grillet. The first ever Australian varietal Chardonnay was produced in 1971 by Murray Tyrrell in the Hunter Valley. New World countries were of small importance to the UK and other European markets, consuming most of their limited production domestically.

Today, the *cru classés* are more successful than ever, but it is unlikely that any of the richer, less tannic wines now produced will last for 50 years. For everyday-drinking wines, grape varieties and brands rule the wine roost. Australia, having long overtaken France, is at the time of writing the number one supplier to the UK off-sales market by volume and value. However, in a falling market, UK sales of Australian wine declined by some 20% in the years 2008–2014. The vast majority of Australian wines on sale in the UK bear no more precise a regional origin than that of the 'super-region' South-Eastern Australia. This spreads some 1250 miles (2000 km) from east to west, encompasses all the wine-producing regions of New South Wales, Victoria and South Australia, and comprises approximately 95% of Australia's production. In South-Eastern Australia, grapes, must and bulk wine may be transported many hundreds of miles, giving the large producers ample components to make their multi-regional blends. The fruit is ripe and the wines correctly made. Within this vast ocean of fruit and variety-driven wines, there are many that are good and some that are very good, but by definition, they lack a sense of place. Australia has, of course, many regions that have that sense of place and can produce wines that are site driven. The most southerly region in South Australia is Coonawarra, relatively cool and at its heart a famous strip of *terra rossa* soil. Well-made, classic wines from Coonawarra tell you that this is where they are from. Cabernet Sauvignon particularly is distinctive, exuding purity and showing expression, with savoury, minty, earthy, dusty and edgy characteristics. The Semillons from Margaret River in Western Australia exude a grassiness that is as distinctive as it is exhilarating.

23.3 The impact of climate upon quality wine production

As we saw in Chapter 2, the climate of the wine region has a considerable impact on the style and quality of the grapes produced and thus the wine made. In cooler regions, grapes may struggle to ripen, and these regions can also be subject to a considerable annual variation in weather patterns, resulting in a huge variation in the style and quality of the harvest. Annual variation in quantity may be substantial, and this is usually a result of the year's (and sometimes the preceding year's) weather, including reduced crops as a consequence of frost, hail or drought. If there is not enough sunshine and heat in the mid to late summer, red grapes will have weak concentration, grassy flavours, green tannins and an acidity that can be swingeing.

Hot regions too can present climatic problems to the producer. Grapes may ripen very quickly, giving high sugar levels (and high pH) but without having had sufficient time for flavour development, or conversely with extreme heat the flavours getting burnt out. Grapes can be affected by sunburn – one way to help counter this is by leaving a denser leaf canopy on the western side of the vines, to shade the grapes from the hotter, afternoon sun. If temperatures exceed 38°C, sugar accumulation in grapes more or less stops because the vine is using all of its energy for staying alive. Provided the vine has enough water, photosynthesis will continue, but the vine's metabolism speeds up with higher temperatures, faster than the rate of increase in photosynthesis. If the temperatures exceed 43°C, there is a risk of the vines dying. In the Peel region, close to Perth in Western Australia, temperatures peaked on Boxing Day 2007 at 45°C, devastating both the quantity and quality of the 2008 harvest.

Producers in regions that are situated on the viticultural edge where, given a warm summer and fine autumn, there is enough sunshine to slowly ripen the grapes with a lengthy hang time on the vine claim that these conditions not only maximise flavours but give naturally balanced fruit. Similar statements are made by those in climates with a large diurnal temperature range – the cool nights lock in colour, acidity and flavour. It should be noted that large diurnal temperature ranges are not restricted to continental climates: Napa, for example, has a maritime climate with relatively small differences between summer and winter temperatures, but high diurnal ranges. The individual mesoclimate, the local climate within a particular vineyard or part of it, also impacts on the quality and style of fruit produced. The *Grands Crus* vineyards of the Côte d'Or in Burgundy are situated in various parts (depending partially on aspect) of the middle of the narrow slope. Towards the top of the slope, it is too cool and windy, and near the bottom, the soils are less well drained. The microclimate, the precise climate within the canopy of a block of vines, will be influenced by many factors, such as row orientation, aspect, surrounding vegetation, cover crops, mulch and so on – the list is endless!

23.4 The role of soils

As we saw in Chapter 3, the physical, chemical and biological properties of the soil, both top- and subsoil, impact on the style and quality of the finished wines. The most important physical characteristics of different soil types are those that govern water supply, water holding and drainage – put simply, quality wine is not produced from poorly drained vineyards. Soil texture depends upon the proportions of gravel, sand, silt and clay, and these affect the vine's ability to take up water, nutrients and minerals. However, the microbiome of the vineyard soil also influences the vine's uptake of these, its disease resistance and the individual aroma and taste characteristics of grapes grown in a particular site. Bacteria from soil can also migrate to grape clusters, and the resulting organoleptic influence is the subject of current research. It is important that soil compaction be avoided, as this results in the soil being deprived of oxygen, and the ability of the roots to grow will be restricted.

23.5 Terroir

The French word 'terroir' defies simple translation. It encompasses the type and quality of soil of a vineyard site, together with the mesoclimate of the area and topographical factors. Mankind, too, has modified terroir by a history of viticultural practices including reshaping vineyards, such as making (and removing) terraces, the installation of drainage systems, and changing the chemical and biological profile of the soil by the addition of chemicals and organic matter.

The soil type and drainage are key components in the concept. Historically, it was generally believed that poor (low in nitrogen), well-drained soils were best for vine growing. However, recent research in the Bordeaux region indicates that a relatively high nitrogen content will increase the aromas of aromatic varieties such as Sauvignon Blanc. Certain soil types are regarded in some countries as inferior to others. For example, in France, both sand and alluvial soils are generally thought not to be conducive to producing very high-quality wine – alluvial soils in particular may be omitted from the areas delimited for 'superior' appellations. At Château Cheval Blanc (Premier Grand Cru Classé A) in Saint-Émilion, the grapes from vines situated on the sandy soils, never go into the Grand Vin. Producers in, for example, parts of New Zealand and Argentina might strongly disagree that such soils cannot give grapes for great wines. However, in France, other soil types remain generally favoured, including gravel, granite, clay and limestone and particularly the *argilo calcaire* mix of the last two.

Professor Gerard Seguin of the University of Bordeaux carried out a landmark study into the chemical properties of soils around Bordeaux, including the top Grands Crus in the Médoc, and the lesser properties. The research revealed that the top châteaux have soils with a high percentage of acidic gravel and pebbles, some 50–62% as compared with only 35–45% in the lesser crus. The top châteaux have soils that are naturally poor in nutrients, and particularly the first metre of depth has a modified chemical profile. The soils are deficient in magnesium owing to high levels of potassium. Seguin considered that this imbalance, with the resultant ion antagonism effect, lowers vine vigour and yields. He states that the finest wines, even within the same property, come from such parcels. The soils are also high in phosphoric acid and organic matter. The clay contents of the soils have been raised, and the sand content decreased. All these modifications can be attributed to the owners being able to afford, over many decades, manure, compost and other amendments that improved soil structure, permeability and resistance to erosion.

The physical properties of the soil are crucial to making great terroir, particularly the drainage and the amount of water stored in the soil at the depth to which roots penetrate. The best soils are free draining, and the water supply to the vines is regulated. The roots of the vine penetrate deep and assimilate the minerals and trace elements found at greater depths. After a period of heavy rainfall, good drainage is particularly important, as waterlogged soils result in the vine taking water via the surface roots and also encourage excessive vigour and attacks of mildew. Seguin's research into the Grands Crus Classés of the Haut-Médoc revealed that the level of their water table drops as the growing season progresses. From the time of veraison in August, when the grapes change colour, and ripening commences, the water table is too low to supply the vines. Thus, during the crucial late-summer ripening period, unless there is rainfall, the energy of the plant goes into ripening the fruit rather than producing vegetation – in other words, the vines are subject to slight hydric stress.

It is only in the last couple of decades that producers in the New World have really begun to discuss their terroirs. The concept was, and for many producers still is, regarded as an Old World marketing gimmick and something that could be used as an excuse for poor winemaking. To a generation of New World producers, armed with the latest research from UC Davis or Waite, many of the expounded terroir characteristics of Old World wines were attributable to poor oenological practices or contaminations: minerality was down to excessive sulfur dioxide, and leather and horsey tones were nothing more than *Brettanomyces*. A few favoured areas in the New World were extolled as being individually superior, a good example being Coonawarra in South Australia, with its famed *terra rossa* strip of soil. Incidentally, such importance was attached to the terroir of Coonawarra (and resultant marketing advantage, wine price and land price) that it has been the subject of a very protracted boundary dispute. However, there are three reasons that the terroir concept remains less important in many New World regions:

1 Irrigation – although this is essential in areas of insufficient natural rainfall, as we have seen above, expressions of terroir are reduced when the vine has an easy supply of surface water. Drip irrigation is efficient but can result in a compact, shallow root zone, with the vine absorbing mineral ions and nutrients from recently applied soil dressings. Deep-rooted vines will absorb these from the soil's natural supply. It is perhaps worth noting here that *V. riparia* rootstocks (and hybrids thereof) tend to be shallow rooting. It cannot be said that irrigation is a bad thing, but it needs careful control. Some producers in Argentina now use both drip and flood irrigation in the same vineyard – the flood irrigation encourages a more expansive root zone. We will look at the concept of inducing moderate hydric stress in the vines in Chapter 25.

2 At many sites, there is only a generation or two of winemaking experience, contrasted with 2000 years or so in the classic regions of Europe. There has been considerable growth of the amount of land planted for wine grape production in the last 25 years in the major New World producing countries. For example, in Australia, the amount of land in production in 1990 was just 42,000 ha, and by 2011 it had risen to 154,000 ha, although it has fallen to approximately 135,000 ha in 2015.

3 Many New World vineyards are massive. For example, the Molina vineyard of Viña San Pedro in Chile's Curicó Valley comprises 1200 contiguous hectares. In Burgundy, the seven delimited, named Chablis Grand Crus vineyards, discussed below, total 102.9 ha. Part of the Molina vineyards is shown in Figure 23.1, and 'La Moutonne,' which is part of the Chablis Grand Cru vineyards of Preuses and Vaudésir, is shown in Figure 23.2. When large vineyards in the New World are machine harvested, the different terroirs, if they exist, are often not separated, and the grapes have no chance of separate vinification.

Figure 23.1 Part of the Molina vineyard in Curicó, Chile.

Figure 23.2 'La Moutonne', part of the Chablis Grand Cru vineyards of Preuses and Vaudésir.

There are many New World producers who take terroir to heart. Philip Woollaston of Woollaston Estate in Nelson, New Zealand, says: 'I believe the place is the most important influence on quality and style. A wine is defined by place, not by variety and not by brand. A brand is nothing. Place is defined, and cannot change. Brands come and go.' In the nearby region of Marlborough, Saint Clair Estate produces a 'Pioneer Block' range of wines, individually vinified, bottled and marketed by the number of the vineyard block. The contrast between these wines made using identical techniques is remarkable. On a somewhat larger scale, New Zealand's largest producer, Pernod Ricard, produces and markets a 'Terroir Series' range of wines defined by the terroirs of Marlborough's Awatere and Wairau valleys.

There is also a move in many other New World countries towards terroir-driven wines, particularly at premium and super-premium price levels. José and Sebastion Zuccardi in Argentina have introduced a range from Valle de Uco showing the distinctive characteristics of the individual vineyard parcels. Finca Piedra Infinita, situated at Altamira, La Consulta, at an altitude of 1100 m (3608 ft), has alluvial soil with large deposits of calcareous rock. Close by is Finca Canal Uco, at the same altitude but with deeper alluvial soils. The wines produced from each, vinified in the same way, are markedly different in taste profile, each exuding the influence of its terroir.

There are many in the wine industry who challenge the concept of terroir. Mike Paul, wine consultant and former brand manager for Penfolds and Lindemans, believes that the New World wine revolution of the 1990s saw terroir diminish in importance and that 'Australia has proved that the wine-maker is the hero, and terroir is just one ingredient in the mix'. However, a tasting of several wines from the same vintage, made from grapes of the same variety, clone and vine age, grown using identical viticultural techniques and produced in the same cellar using identical winemaking techniques, can reveal that terroir has a profound influence on the flavours of the finished wine. For example, in the district of Chablis, the seven named Grand Cru vineyards lie contiguously in a hillside arc just north of the village. The elevation and aspect vary somewhat, and the soils, basically fossil-rich Kimmeridgian clay, also have subtle differences. The wine from each of the vineyards tastes different, even when the vinification is identical. Table 23.1 shows the named Chablis Grand Crus vineyards and indicates the differences in the flavours of wines produced therefrom.

Barone Francesco Ricasoli, producer of some very fine individual vineyard Chianti Classicos, states: 'I am not producing appellations, or varietals. I am producing terroir wines'. Soil mapping of the Ricasoli Brolio vineyard shows there are 19 different soil types, with five main geological formations. The altitude ranges from 250 to 450 m. The produce of each plot is harvested and vinified separately in order to respect the requirements and integrity of each terroir, a process that is replicated by terroir-driven producers around the world .

Table 23.1 Chablis Grand Crus

Chablis Grands Crus	
The Chablis Grand Cru vineyards cover a total area of 100 ha (the size of just one of the larger Médoc châteaux) and comprise seven named AC Grand Cru Vineyards:	
Vineyard name	**Size and styles of wine**
Bougros	12.6 ha. The vineyard is situated at the western end of the arc of the Grands Crus and produces wines with a vibrant but solid, masculine character
Preuses	11.4 ha. Producing wines with a succulent almost exotic, ripeness, fleshy, richer and fatter than the other Grands Crus
Vaudésir	14.7 ha. Ripe wines with an underlying mineral and spicy character, feminine and very complex
Grenouilles	9.3 ha. Softer, delicate wines that are very fruity, racy and elegant
Valmur	13.2 ha. Wines that show delicate floral aromas and are perhaps the softest of the Grands Crus
Les Clos	26 ha. The largest of the Grands Crus, and considered by many as the finest, producing wines that are rich and powerful yet steely, with exhilarating minerality
Blanchot	12.7 ha. Wines that show floral and mineral tones and are of delicate structure; they can be firm when young, but develop wonderful complexity

23.6 The vintage factor

As we have seen, each vintage in any region or even vineyard will vary in quantity and style as well as *quality*. Cold weather early in the season, particularly frost at budding time and in the weeks that follow or windy weather at flowering time, can result in a considerably reduced crop. Hail in late summer can devastate a crop and even the vines. Such problems are not exclusive to Europe. Many New World regions are at risk, e.g. Mendoza, Argentina's largest wine region, can be victim to cold, frost, wind and hail. Plantings on slopes are generally at less risk of frost, as the dense, cold air flows down the slopes. Of course, the reduction in yield by some weather factors may result in an increase in quality, but the key to this is that the limited fruit production must be clean and balanced.

While plenty of summer sunshine is clearly desirable, if it gets too hot the quality of fruit will diminish. Exposed grapes can become sunburned and suffer berry crinkle resulting in dried fruit and cooked flavours in the wine. The balance between acidity, sugars and phenols is crucial for quality wine production. At temperatures above 38°C, the vines' metabolism can shut down, resulting in under-ripe fruit. A little rain in the spring or early summer is generally desirable, for if the vines are too stressed, quality will also be reduced. Perhaps most desirable of all is a period of dry, warm sunny weather in the weeks prior to harvesting. If the summer has been cool, and the grapes are healthy but under-ripe, such conditions can lead to a high-quality vintage. This was the case in Bordeaux in 2014, when a warm and dry September helped turn a harvest that would have been uninspiring into a classic vintage. However, many otherwise great vintages have been wrecked by autumn rains arriving while the fruit is still on the vine. Vintages vary in style from the big, rich wines produced in years when it has been hot, to the lighter but very good wines produced in years with less sunshine. Grape varieties ripen at different times; for example, Merlot ripens up to 10 days before Cabernet Sauvignon. Accordingly, in Bordeaux, for instance, if the rains come in late September, the Merlots will probably already be fermenting, and wines dominated by this variety may be excellent, while the Cabernet Sauvignon-dominated wines may be less good. On the other hand, if August and early September have been cool with excessive rain, but then there is an Indian summer later in the month and into October, it is the wines majoring on Cabernet Sauvignon that may excel.

Rain at harvest time can be a nightmare, creating a mood of depression throughout a wine region. *Botrytis cinerea*, as grey rot, can become established on grapes very quickly. The fungus thrives in a humid atmosphere, and infected grapes that have taken up water from the soil will expand, resulting in those in the middle of the bunch becoming compacted and damaged. Bunch thinning can help to reduce this problem. If fruit is delivered to the winery wet, it will dilute the juice. Mechanical harvesters can prove their worth here, for if rain is forecast, they can bring in the crop very quickly. If a short period of rain is forecast, provided the fruit is healthy, a grower may take a gamble and delay picking until afterwards, hoping for an Indian summer to achieve more ripeness. If hand harvesting takes place, pickers may be instructed to leave damaged or rotten fruit on the ground. Effective sorting of the harvest crop, as detailed in Chapter 25, is essential in wet harvests.

The vintage factor adds a sense of time to that of place. The variation in style, detailed flavours and quality, reflecting the growing year, adds another layer of individuality to the wines of a particular region or producer. This is unlike every other drink, where style and consistency from batch to batch are the norm. A vertical tasting, that is tasting a number of vintages of the 'same' wine (albeit at different stages of maturity), is invariably one of the most fascinating and revealing of wine-tasting experiences, because of the variation in product characteristics that such a tasting reveals.

CHAPTER 24

Constraints upon quality wine production

'Mother nature isn't always kind' is the oft-heard lament of those working with vines and their produce, basically an agricultural product turned by art and science into the most talked of beverages, 'but it's much kinder than the government, the supermarkets and the bank'. In other words, it is often the business and economics of producing wine that can put more constraints on quality than natural forces. In this chapter, we consider the financial, skills, legal and environmental pressures that may constrain wine quality.

24.1 Financial

The number one aim of any wine producer, grower or winemaker is to make a profit. Indeed, it can be argued that for corporations large and small, maximising profits is vital to attract further investment or other finance, whether it is from shareholders or financial institutions. It might be hoped that such further finance is invested, at least in part, into improving both the individuality and quality of the product, both of which should give a competitive edge. However, all too often, it is the marketing and promotional budgets that are expanded.

Increases in profit may be achieved by the following:

1 Increasing prices – this may involve 'repositioning' the product. The price charged cannot, of course, be more than the market will bear.
2 Increasing quality – this may boost sales or lead to (1) but will usually also involve an increase in costs, although it can often be achieved simply by improving practices.
3 Increasing production and sales – this may result in reduced quality, and it may be the case that the market will not swallow such an increase without downward pressure on price.

Wine Production and Quality, Second Edition. Keith Grainger and Hazel Tattersall.
© 2016 John Wiley & Sons, Ltd. Published 2016 by John Wiley & Sons, Ltd.

4 Increasing the market's perceptions of the product, which will probably involve more spend on marketing but may lead to (1).

5 Reducing costs – this may result in reduced quality as discussed below, but production-cost reductions can be achieved by greater efficiency.

24.1.1 Financial constraints upon the grower

A grower, who is not also a wine producer, will sell grapes either to a cooperative of which he or she is a member, or directly or indirectly to a wine-producing company. The price paid will depend on many factors. These include contractual agreements, the state of the market, the variety grown and the quality and sanitary condition of the crop.

Growers may have long-term contracts with wine producers, often requiring them to adhere to stipulated practices and to produce a crop to the buyer's specification. It is not unknown for buyers to renege on such contracts in times of financial difficulty or glut, and if the grower is tempted to seek legal address, they will probably be bankrupt by the time the case is heard. However, without such contracts, the grower is at the mercy of the market, and this can prove disastrous. Cooperative members, too, are exposed when market prices fall.

In 2014, according to figures from the Winemakers Federation of Australia, just 12% of the country's wine-grape growers made a profit. In some areas, the figure was even less: in South Australia's Riverland (home to Berri Estates, the largest winery in the southern hemisphere), just 1% of growers reported a profit. The average price Australian growers achieved for their grapes in 2014 was AUS$441 per tonne. Since 2001, when the price peaked at nearly AUS$933 per tonne, prices have been generally falling, with the exception of 2008, when they rose following drought conditions. The ultra-low average per tonne prices obtained in 2014, and the fall from the already low figures of 2009, for some varieties in the inland regions is quite shocking: Chardonnay from Murray Darling $220 ($308 in 2009), Colombard in Riverina $157 ($197 in 2009) and Chenin Blanc in Riverland $187 ($295 in 2009). Perhaps unsurprisingly, growers in areas famed for their quality achieved much higher prices with Pinot Noir from Tasmania commanding the highest price in 2014, an average of $2672 per tonne, with some growers achieving over $3500. All these prices were obtained from the Australian Grape and Wine Authority, and are in Australian dollars. A graph showing the average price obtained for grapes in Australia in each year from 2000 to 2014 is shown in Figure 24.1. The reader may wish to contrast the above grape prices with those obtained by growers in the Champagne region in 2014: between €5170 and €6060 per tonne (depending on the village the grapes were grown), which equated to between approximately AUS$7755 and AUS$9090 per tonne!

The losses made by most growers in Australia have been particularly severe during the last decade, there have been many bankruptcies, and the

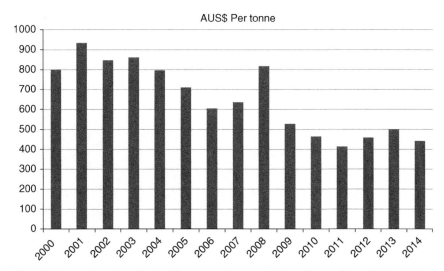

Figure 24.1 Average price obtained for grapes in Australia in each year from 2000 to 2014.

uprooting of vines in some areas continues. The situation in some other countries is little better. In South Africa, according to figures from S A Wine Industry Information & Systems NPC (SAWIS), the average price growers obtained per tonne in 2013 was ZAR3800 (approximately AUS$407 at the then exchange rate), which was up from a low point in 2007 of ZAR2781. However, during the period 2008–2013, inflation (Retail Price Index) in South Africa totalled 57%, meaning that inflation-adjusted income was actually substantially down on the 2007 low point.

Of course, growers blame the purchasing wineries for paying low prices and driving them to desperation; the wineries blame their customers, particularly the large supermarket groups both locally and on key export markets. Exports account for 60% of Australia's wine sales by volume with 36% of the exports destined for the UK market (source: Wine Export Approval Report, December 2014; Wine Australia, 2015). South Africa exports 57% of production, with over 21% destined for UK and 19% to Germany (source: SAWIS). The supermarket buyers in both these countries are noted for being particularly aggressive. Of course, growers in many European regions, too, have felt the consequences of the relentless squeeze on prices, and with the EU redirection of production subsidies into export marketing budgets, incomes have further diminished.

In times of financial pressure, savings have to be made. A reduction in labour costs will see an instant improvement in the monthly cash flow but a decline in the quality of the vineyard, either rapid or slow. With less spending on preparations, soil adjustments and pest and disease control, the quality of

Figure 24.2 Untended vines running riot.

the harvest may be compromised. If total income is down, the temptation is to increase yield in order to have more product to sell, thus to maintain cash flow. This may perhaps be achieved by less hard pruning, increasing irrigation and abandonment of green harvesting. Such yield increases may further reduce quality (and result in a further reduction in prices obtained). In some cases, vineyard work, other than harvesting (by hand), stops altogether, with a disastrous impact upon quality. An untended vineyard (in southern France) is shown in Figure 24.2.

24.1.2 Financial constraints upon the winemaker

The relentless pressure on producers from retailers striving to maintain price points, offer promotions and improve margins means that quality compromises often have to be made in the winery. The market rules and costs must be made to fit. Fruit selection may be minimal or non-existent, as witnessed at some large wineries by the 10-tonne tipper trucks discharging into the receiving hoppers that directly feed the crushers. An illustration of this taking place in Chile is shown in Figure 24.3. Fermentations may be undertaken warmer (chilling is expensive in energy). Indeed, at every stage, the must and wine will be speeded on its progress through the winery, for time is money. This will involve rapid clarification and stabilisation techniques that some critics consider may compromise quality, e.g. tangential filtration and flash pasteurisation. One key area where costs can be contained is in the 'oaking' of the wine.

Figure 24.3 Ten-tonne truck discharging loose-filled grapes.

24.1.2.1 Oak treatments as a substitute for barrel ageing

Purchasing new barrels is expensive, and buying used barrels is risky, particularly with regard to possible contamination with *Brettanomyces*, acetic bacteria or, disastrously, haloanisoles. Of course, the cost of barrel-maturing wines is not just the capital costs of the barrels but the increased amount of labour required and the loss of wine owing to evaporation. The high cost of barrel maturation is prohibitive for wines with inexpensive price points, so numerous alternative ways of oaking wines have been developed. These include oak powder, granules, chips, rods, ladders and staves. The rates of usage and approximate costs per litre of wine for the various products are shown in Table 24.1.

The total cost per litre of wine aged using chips or (usage rate 1.5–2 g/l) will be less than one tenth of a wine aged for one year in an equal mix of new, one-year-old and two-year-old barrels – perhaps €0.10 per litre contrasted with €1.2 per litre.

Of course, when wine is maturing in oak barrels, there is an interplay between the wine, the wood and the atmosphere: wine evaporates through the pores of the wood, and a controlled oxygenation takes place, together with the absorption of oak products. This does not happen when the wood is on the inside rather than the outside, such as the use of inner staves in a tank.

Table 24.1 Rate of usage and approximate cost per litre of wine for various oak products

Product	Typical dimensions	Usage (approximate)	Unit price	Cost per litre of wine
Large tank Staves French	2200 × 100 × 12 mm	200–250 litres of wine per stave	£17.50	£0.07+
Oak barrel Chain French	240 × 22 × 7 mm	1 per 3rd or 4th fill barrel	£21.80	£0.097
Small tank Staves French	960 × 50 × 7 mm	35–40 litres of wine per stave	£3.80	£0.095
Oak cubes French	50 × 50 × 7 mm 2.5 m² per bag	1200 litres of wine per bag	£75.25 per bag	£0.062
Oak chips French	7–10 mm	1.5–2 g/l	£6.27/kg	£0.0125
Oak chips American	7–10 mm	1.5–2 g/l	£2.09/kg	£0.004
Oak granules French	2–3 mm	1–1.5 g/l	£6.95/kg	£0.010
(Oak powder used with fermentation)	Powder	600 g per tonne of grapes	£4.15/kg	£0.003

Micro-oxygenation, by which minute amounts of the gas are bubbled into wine in tank, is one way of attempting to emulate the breathing of wine in barrel. However, the use of oak chips, granules, shavings or powder may be a poor substitute for ageing in oak barrels, the products often imparting incongruous sweet and sappy flavours and, sometimes, harsh textures. Advocates of using alternative oak products combined with micro-oxygenation claim that the process is not necessarily inferior to barrel ageing, stating a better control of the amount of oak added, and flexibility of usage, including the timing of additions just when required. We will return to the topic of barrel ageing in the next chapter.

24.2 Skills and diligence

It is easy to imagine that in the high-tech world of twenty-first century wine production, all involved are well educated and highly skilled in matters relating to their particular field of work. This is certainly the case at good-quality producers, where the price commanded for the wine warrants, indeed necessitates, that every aspect be performed meticulously. At lower levels, lack of training, skills and diligence is often all too apparent. Vineyard work in particular is often carried out by contractors who employ low-paid operatives who have had the bare minimum of training, including immigrant labour hired on a day-by-day basis. Without any sense of investment or ownership in the business and detached from the finished product, workers underperform. Tasks may be performed by operatives who know what they are doing but not why they are doing them. A typical full-time vineyard worker in South Africa earns ZAR620 per week (approximately £34.00 at the rate of exchange at the time of writing). It is little wonder that some of the vineyards exhibit the lowest standard of viticultural work the authors have seen anywhere in the wine world, an opinion endorsed by other wine writers. Pruning is a task that requires skill and a 'feel' for the vine as a living being, but this task is often undertaken by pieceworkers who will work, at best, according to the simple formula instruction they have been given, e.g. to leave 12 nodes. Good fruit only comes from balanced vines, and sensitive pruning is one of the necessities to achieving balance. It is not uncommon when walking in New World vineyards to find irrigation pipes that have become blocked or burst, resulting in some areas that are flooded with nearby land and vines parched.

Skills and diligence in the winery, too, are of paramount importance. Even the most knowledgeable and fastidious winemaker will not be present at all operations in a large winery and the operatives, 'cellar rats' and interns may compromise quality owing to ignorance, carelessness or laziness. Cleanliness is crucial, but chlorine-based cleaning agents should be avoided at all costs, to avoid the risk of haloanisole contamination. If a destemmer-crusher is out of adjustment or not operated carefully, problems can ensue. For example, it is important that if a destemmer is used, the equipment should remove and eject the entire stems. Poor equipment may break up the stems, giving a wine that is very stalky and harsh as a result of the presence of stem tannins. If the rollers of the crusher are out of adjustment, the pressure may be too great, crushing the grape pips and releasing bitter oils to the must.

Keeping tanks completely full, for example by placing a floating lid directly on the surface of the wine, is vital, otherwise surface film yeast, bacteria, acetic acid or oxidation will result. Barrels must be clean and sterilised between uses. The general belief is that wine must be topped up regularly during the first year of maturation in barrel, but some winemakers challenge this, with the proviso that the bung is in tight, and the barrel has been rolled to the 2 o'clock position. All would accept that oxidation is the great enemy. Filters must be regularly flushed out, and the filter material replaced as necessary. Poor filtration can lead to cardboard-like flavours in the wine. It is important that the bottling process take place quickly. As vats are emptied, bottlers should blanket the wine with inert gas, in order to prevent oxidation. Although bottles arrive at the plant shrink-wrapped on pallets, they should again be washed before use. Haloanisoles (discussed in Chapter 21) have been found in pallets of supposedly clean, new bottles arriving at bottling lines which, even at low levels of contamination, will strip the bottled wine of fruit.

24.3 Legal

It may seem strange that legal considerations could constrain quality, particularly within the EU, until it is understood that many of the viticultural and oenological practices stipulated in the wine regulations were agreed after consulting, and perhaps under some pressure from, the large producers. It can be argued that the control of production costs, and thus control of production methods, was the prime driving force behind the regulations. As we saw in Chapter 22, practices that are not in the regulations as being permitted are forbidden. Small producers, in particular, often find that working within such a straightjacket of legislation restricts their ideas of creating individuality and improving quality, at least if they want to sell their product as PDO wines. Accordingly, some producers chose not to sell their wines as such, but to label as a Protected Geographical Indication or simply as 'wine', e.g. Vin de France. However, for fear of losing an illustrious and valuable PDO status (e.g. Hermitage AOP), other producers do take steps to 'standardise' their wine, to ensure that it passes the tests of compliance, technical analysis and *typicité* at the tasting.

24.4 Environmental

In a wine world where priority is now being given to measures that will limit damage to, and if possible enhance, the environment, it may seem anathema that such considerations can have a negative impact on quality. Certainly, many environmental considerations have had positive effects. During the 1960s, 1970s and 1980s, it seemed that one could not enter a vineyard without seeing routine preventative spraying or applications of 'yield-enhancing' chemicals. Now, a great many quality-conscious and environmentally friendly growers practise 'integrated pest management', and the last two decades have seen a remarkable increase in the number of organic and biodynamic wine producers worldwide. Additionally, cover crops abound, and the gentle hum of 'beneficial' insects is the main sound to be heard in the vineyard! Significantly, the standards of vineyard hygiene on the quality-conscious estate are higher than ever, the vines are better balanced, and the quality of fruit has improved.

However, today, the buzzwords are 'reducing environmental impact', 'reducing the carbon footprint' and being a 'carbon neutral producer', and retailers and distributors are increasingly demanding that suppliers act to achieve this. It does, of course, provide very good public-relations material, even if carbon-neutral status is achieved by simply buying carbon credits. Traditional filtration materials such as kieselguhr and filter sheets are effective but becoming less popular because their residue goes to landfill. There is also a quality price to pay for demands to move to bottling or otherwise packaging wines in the country of sale rather than the region of production, and in lighter-weight containers rather than glass bottles.

Historically, the shipping of wine in bulk was the norm, and bottling would take place by merchants in the country of destination. Wines have been shipped in barrels for centuries, but during the twentieth century other means of transport began to be utilised, including, for cheaper wines, ships' tanks, SAFRAP (lined mild steel) containers and road tankers. Of course, producers had to prepare, adjust and protect wines for the journey, and the bottler would have to prepare the wines for bottling. There is no doubt that wines would often deteriorate during transport, and the integrity of the product was often compromised by the bottling merchant. So, in the second half of the twentieth century, there was a growing move, especially for fine wines, towards bottling at least in the region of origin and at best at the property of production. Compulsory bottling at the château for *cru classé* Bordeaux wines was enacted from the 1972 vintage, although most chateaux had been bottling all or part of their production at the property for decades. Today, certain AOP (AC) and other national equivalent wines must, by law, be bottled in the region of production.

A case of 12 × 75 cl bottles of still wine weighs approximately 15–16 kg, with 9 kg being the weight of the liquid and the rest being that of bottles, packaging, etc. There is also much 'wasted' space within the case, so shipping in bulk for bottling at destination, thus reducing both the weight and space required, makes prima facie environmental sense. As we have seen, wines transported in bulk need to be stabilised for the journey (which may include an extra dose of sulfur dioxide) and, after arrival, will need further treatments and additions to prepare for bottling. Most quality-minded producers and nearly all critics believe that minimal handling is a key to high quality. The moving of bulk wines, especially delicate whites, may result in a loss of freshness as is witnessed by numerous examples of Chablis bottled not in the district but further south in the Côte d'Or.

The bottling of some inexpensive wines in lightweight plastic bottles that have something of the feel of their glass counterparts was introduced over a decade ago. A glass bottle weighs approximately 400 g, and the plastic bottles just 54 g. Without doubt, plastic bottles compromise quality, for, unlike glass, which is a totally impermeable and inert material, plastic allows a small amount of oxygen ingress. The material also exhibits flavour absorption and wine acids can attack plastics. Sulfur dioxide levels have to be increased for wines bottled in plastic, with a resulting impact of a loss of fruit. The shelf-life of the wine in plastic bottles is also reduced. As a result of these issues, together with negative consumer perceptions, these containers have had very limited success in the marketplace, in spite of strident efforts by the plastics industry to promote the 'environmental benefits' of the packaging. The introduction of lighter-weight glass bottles is perhaps a more positive move, although there are some issues regarding their strength and, in some cases, the reduced degree of filtering of damaging UV light.

CHAPTER 25

Production of quality wines

In this chapter, we will consider some of the key factors in the production of quality wines. Achieving quality, in what is basically a natural product, requires careful attention at every stage from planning the vineyard and winery, right through to bottling and storing the product. Fundamentals are crucial, but so is attention to detail. This chapter explains key concepts to achieve quality, but the list of topics covered is far from exhaustive.

25.1 Yield in vineyard

The subject of yield, permitted and actual, and its impact on wine quality is a subject that arouses much discussion. The conventional view, particularly prevalent in the Old World, is that restricting yield enhances the quality of the fruit produced. As we have seen, this concept is embodied in the laws of AOP – the 'higher' the appellation, the more restricted the yield.

Yields are usually expressed in hectolitres per hectare (or per acre) or, particularly in Australasia, in tonnes per acre. Examples of basic permitted yields in the Bordeaux region are shown in Table 25.1. It should be noted that the average yield in the region as a whole in the years 2005–2014 was 48 hl/ha.

There are several ways for growers to restrict yields, including the density of planting, winter pruning, stressing the vines by deficit irrigation in countries where irrigation is practised and green harvesting. However, forcing low yields can be detrimental to quality, particularly so in vigorous sites, but also in vineyards where vine balance is less of a problem. Frédéric Bonnaffous, Estates Director of the Dourthe portfolio of well-performing vineyards in Bordeaux, says: 'the vines must be in balance. If you restrict the yield to 20 hl/ha the

Wine Production and Quality, Second Edition. Keith Grainger and Hazel Tattersall.
© 2016 John Wiley & Sons, Ltd. Published 2016 by John Wiley & Sons, Ltd.

Table 25.1 Examples of basic permitted yields for Bordeaux AOPs

	Yield (hl/ha)	Yield in 2014 (hl/ha)
Bordeaux AOP – white (region)	67	65
Bordeaux AOP Rosé et Clairet (region)	62	60
Bordeaux AOP – red (region)	60	57
Bordeaux Supérieur AOP – red (region)	50	51
Haut Médoc AOP (district)	48	
Margaux AOP (commune)	45	
Sauternes AOP (commune)	25	

The basic individual yield stated (*rendement de basse*) may be varied each year (*rendement annuel*) by Institut National de l'Origine et de la Qualité.

wines will be so intense that there will be no expression of terroir'. That said, there is no universal agreement as to what is the optimum yield for the production of high-quality wines, even within individual districts, and in any case most growers have to strike a balance between quality and economic production levels. Some growers will aim at the maximum possible permitted or achievable yield without concern as to the negative impact on quality. In Australia, unless top-quality wines are the objective, most growers will state that any yield below approximately 4 tonnes per acre (approximately 10 tonnes/ha) is uneconomic for most varieties. For red wines, this equates to a yield of approximately 60–72 hl/ha, amounts in excess of the legally permitted yield of even the most basic (regional) Burgundy appellations.

The yield per vine is as important as the yield per hectare. A grower might claim a low yield, but if many of the vines are dead or dying, the yield per productive vine might be relatively high. In France, the rules of AOP provide that if the amount of dead vines comprises less than 20% of the total, the normal permitted yield per hectare applies; if the number is greater than 20%, the permitted yield is correspondingly reduced.

25.2 Density of planting

The more vines per hectare, the more each vine competes for natural resources, particularly water. With limited amounts of rainfall, or deficit irrigation, the vines are slightly stressed. Vigour is restricted, and the energy of the plants goes into ripening a limited quantity of fruit. Having to compete for water encourages vines to send their roots deep, where they pick up minerals and trace elements.

As we saw in Chapter 4, density of planting will depend on several factors, including legal and topographical, orientation as well as other aspects of the vineyard site, labour costs and the use of mechanisation. In parts of Bordeaux, including many vineyards of the Médoc, vines are planted at a density of 10,000 per hectare. In some vineyards of Burgundy, e.g. those of Domaine de la Romanée Conti, the figure rises to 12,000 vines per hectare. As Frédéric Bonnaffous says:

> more vines per hectare, depending of course upon soil conditions, leads to better quality. It's rather like having to take a big weight to the top of a mountain. If you have more people, each will carry less weight and therefore be less tired.

Some small, quality-conscious producers are taking the concept of dense planting even further – the Domaine Léandre-Chevalier estate in Blaye, Bordeaux, has vineyards with up to 33,333 vines per hectare. Mechanical working in densely planted vineyards can pose problems of soil compaction – the wheels of the straddle tractors follow the same path with each pass. Some vignerons have designed special machinery to avoid this or use horses, whose hooves land at a different spot with each passage. Regular tilling puts oxygen back into the soil, or alternatively the use of cover crops including cereals, legumes and grasses helps alleviate the problem of compaction.

At 10,000 vines per hectare, and based on a typical yield in the Haut-Médoc district of Bordeaux of 48 hl/ha, the average yield per vine is 0.48 litres. In some New World countries, the planting density might be as little as 1080 vines per hectare, and a yield of 45 hl/ha would result in an average yield per vine of 4.5 litres. However, yields in some New World countries can be high and historically have been crazily so. There are areas, such as Big Rivers in Australia where, prior to 2006, a yield of over 200 hl/ha was not uncommon, and based on a density of 2000 vine per hectare, the production per vine might have been 10 litres. Today, the yield in Big Rivers is generally in the region of 90–100 hl/ha. To increase the density of plantings, some growers are now planting a further row between two existing rows of vines, thus doubling the number of vines per hectare. An illustration of such plantings, in Central Otago, New Zealand, is shown in Figure 25.1.

25.3 Age of vines

It is generally accepted that young vines do not yield high-quality fruit, although as the vines are often in balance, the first crops of some varieties such as Sauvignon Blanc or even Pinot Noir can produce some classic varietal flavours. It is also generally accepted that old vines produce the finest wine grapes, albeit in a lesser quantity. After an age of approximately 20 years, the vigour of the vines is reduced, and balance is restored. However, there is no consensus as to what age vines become mature and at what age they become old. It is not uncommon to see the words 'vieilles vignes' on labels of French wine, or 'alte reben' on those from Germany, but the terms have no legal definition.

Figure 25.1 Inter-row plantings in New Zealand's South Island.

Interestingly, aromas and flavours most true to the variety generally do not come from old vines, which perhaps give more terroir characteristics. Many top producers, including Bordeaux *crus classés*, exclude the crop taken from young vines from their Grand Vin. Some define 'young' as less than 7 years old, others as less than 10. On the steep slopes of the Dr Loosen Estate in the heart of Germany's Mosel region, Ernie Loosen replaces dead or dying vines on a one-by-one basis and thus maintains a particularly high average vineyard age, a factor that he believes is a major contributor to the acclaimed, exceptional quality wines. Elsewhere, producers might choose to replant entire blocks, as illustrated in Figure 25.2 (photo taken in Saint-Émilion).

25.4 Winter pruning and vine balance

For most growers, this is the main way of restricting yield and preparing for the desired canopy. The more nodes that are left on the vine, the greater the potential yield. The range is six to 40 nodes – those who prune very hard may allow an extra node or two in case there is any die-back, perhaps as a result of frost. Some growers have a fairly rigid formula, e.g. a vine will be pruned to 10 nodes. Others prefer to consider the vines as individuals and decide on an

Figure 25.2 Replanted vineyard in Saint-Émilion.

appropriate number according to the age, health and vigour of each vine. Always, the aim should be to achieve vine balance, i.e. the balance between fruit and vegetation. A simple way to measure this is to weigh the prunings and compare with the weight of fruit yielded by the vine in the previous harvest. The weight of fruit should be between five and 10 times the weight of prunings: more than this and the vine is over-cropping (and should be pruned harder); less than this, then it is too vigorous, and more nodes should be retained. Careful pruning and training (including subsequent summer pruning, removal of laterals and leaf plucking) also help maintain a healthy canopy, reducing the likelihood of cryptogamic diseases such as powdery mildew, also known as oidium, and downy mildew, also known as peronospera. This is especially important if a high number of nodes were retained during winter pruning, leading to a dense canopy.

25.5 Stressing the vines, vine and nutrient balance

Vines, like people, work best when slightly stressed. This is a conventional view that is now challenged. Obviously, a grower has no control over rainfall, but in countries and regions where irrigation is permitted, the amount of water given to vines has a major impact on quality. Growers may choose to practise deficit irrigation. Neutron probes in the soil, or more simply the use of evaporating pans, indicate when water is required. Stressing the vines causes the roots to synthesise abscisic acid, sending this to the leaves and deceiving them into reacting as though there are drought conditions. Shoot growth stops, and all energy goes into ripening the fruit. Moderate water deficit can double or treble the concentration of the precursors of the varietal thiols that will be released during fermentation. Partial root drying is a recent technique. The roots on one side of the vine are given insufficient water, while the other side is irrigated normally. Before any damage occurs to the roots, the irrigation pattern is reversed. However, as with many viticultural matters, the concept of vine stress and deficit irrigation as a vehicle to improve quality is far from universally accepted. For example, Neal Ibotson, of the multi-award winning Saint Clair Estate in the Marlborough region of New Zealand, states that stressed vines give poor quality, a view endorsed (particularly in the case of Sauvignon Blanc) by Kevin Judd, owner of Greywacke wines and former chief winemaker at Cloudy Bay.

Quality is affected not just by whether the vines are stressed but when they are stressed. As we have seen, it is beneficial for Sauvignon not to be stressed, especially if pyrazine aromas are desired in the wine. Unstressed Cabernet Sauvignon also produces very pyrazine-dominated wines. While, with Sauvignon Blanc this is considered desirable, it is not the case with Cabernet Sauvignon, particularly as the tannins are also adversely affected by excess water. It is generally accepted that sustained stress is undesirable, as this results in weak root growth, restricting the vines' ability to take up minerals.

Clearly, it is important for a grower to know the levels of stress the vines are undergoing. An excellent method of determining this is by using a pressure chamber to test the midday leaf water potential (LWP) or stem water potential. An alternative test is used to measure the pre-dawn water potential (PDWP) as the vine is regarded as being in equilibrium at that time of day. Petiole analysis is a vehicle by which a grower may determine if vines have sufficient nutrients and minerals. This may be undertaken twice in a growing season: at full flowering and early veraison. If deficiencies in micro-nutrients such as zinc, manganese or iron are noted, these may be remedied by foliar sprays, usually resulting in an immediate response from the vines. These micro-nutients are often fixed in the soil and unavailable to the plants.

Recent developments, for those who can afford them, include precision viticulture and vineyard mapping, in which aerial photography with multispectral and infrared imaging is used to show areas of vineyards that are lacking water or nutrients. The information obtained is transmitted to the machinery in the vineyards, which then adjusts the additions to individual areas of vines as required. With sophisticated drip irrigation systems, it is even possible to vary the amount of water given to each vine – all controlled by the computer in the vineyard manager's office.

25.6 Green harvesting

This involves removing fruit from the vine before ripening. Having ensured that the vine is bearing at least the desired amount of fruit (having endured any early season climatic, disease and pest risks), a green harvest of fruit in excess of the desired yield will allow the energy of the vine to go into fully ripening a lesser quantity of high-quality fruit. If utilised, the process may take place at or just after veraison, when it is clear which grapes may be left to develop. Advocates believe that flavour as well as sugar concentration is improved. However, many viticulturalists challenge the concept, believing that if green harvesting is necessary, the vines are out of balance, and it is simply a method of artificially reducing the yield. Bruno Prats, former owner of Château Cos d'Estournal, is quoted as saying: 'It's like pushing the gas pedal and the brake at the same time.'

25.7 Harvesting

The timing and method of harvesting have a major impact on quality. A grower or producer's decision as to when to harvest will be decided upon with respect to many factors, including seasonal parameters, the weather forecast, the risk of botrytis, the balance of sugar and acidity in the grapes, the maturity of skin and seed tannins, and the formation of aromas and aroma precursors. Of course, as the harvest cannot be organised at a moment's notice, careful analysis and sophisticated prediction tools are employed.

Historically, especially in some regions of the Old World, many growers would pick early – in fact pick too early, for the grapes often lacked phenolic ripeness or even the desired levels of sugars. Growers also lived in fear of autumn rains spoiling the harvest with bunch rot and dilution, and so undertook early 'insurance' picking. In order to ensure that the grapes have an acceptable, minimum level of ripeness, the rules of AOP (AC) include, in most regions, a *'ban de vendage'*, a date variable each year before which the harvest may not take place. Nowadays, however, growers more often wait for phenolic ripeness, which, in warm and hot climates, may not come until after the desired levels of sugar and acidity are present. This can result in over-alcoholic and over-loud wines. Blair Walter, of Felton Road Winery in New Zealand's Central Otago region, states that he is making a conscious effort to pick earlier and is not afraid of a green edge. Jeff Synnott, formerly of nearby Amisfield

Wine Company and now of Waipara West, agrees: 'Our notion of ripeness and quality has somehow become confused'. Speaking in 2015, Francisco Baettig, head winemaker of Chilean producer Errazuriz, comments:

> Today I'm striving for more aromatics and am not afraid about having leafy notes in Cabernet Sauvignon – a hint of herbaceousness is part of its character… . I'm also picking around 10 days earlier to achieve more freshness but it's not only about picking early. I'm irrigating more as I don't want to stress the vines out too much.

25.7.1 Mechanical harvesting

Some viticulturalists claim that in order to pick the right fruit at the right time, machine harvesting is essential. A machine can pick in one day the quantity of grapes that would require 40–80 manual pickers. The technology of mechanical harvesters has improved considerably during the last 25 years. The harvesting system generally now used is that of bow-rod shakers. The fibreglass rods move the fruiting zone of the vines to the right and then rapidly to the left, and using the principle of acceleration and sudden deceleration of a mass, the grapes are easily snapped off the stalks, which are left as skeletons on the vines. Lightweight berries, including shot berries, raisins and those that have been bird eaten, do not have the mass and so are left behind. The fruit is initially collected on conveyers comprising interconnected shallow baskets, which are moving in a rearwards direction at the same speed as the harvester is moving forwards. MOGS (matters/materials other than grapes) are removed by two methods: spinning blades just above the trays will remove large MOGS, and vacuum systems remove light matters such as leaves. Some of the latest designs of mechanical harvesters are now able to pick whole bunches.

Advocates of mechanical harvesting cite the advantages as follows:
1 Machines work fast, and the grower can utilise them at the time of optimum grape ripeness.
2 The fruit picked is cleaner than may be the case with unskilled and/or piecework labour.
3 Machines pick at a controlled rate, so there should be no delay in processing, which gives an advantage over manual picking which can result in queuing for the sorting tables and destemmer-crusher, which would result in fruit deterioration.
4 Machines can work at night and deliver desirable, cool fruit to the winery.
5 The cost of picking is generally less than half that of using manual labour, although the cost of the machine has to be amortised.

25.7.2 Hand picking

In spite of considerable improvements in recent years, machines do damage grapes to a small extent. Trained pickers can select fruit according to requirements and pick into small bins, sometimes just 12 kg capacity, which will also minimise fruit damage. Hand picking is the only option if the winemaker wishes to vinify separately the product of small parcels of individual terroirs. Hand picking is essential if whole clusters are needed for white wines, and there is a growing trend for red winemakers to use at least some whole clusters in their ferments.

25.8 Delivery of fruit

Ideally, the fruit should be transported to the winery in small containers, so that it does not crush itself under its own weight. Grapes should be processed as soon as they are harvested – some top producers aim to have fruit in the crusher within 1 h of picking. The longer the grapes are left waiting before they are processed, the more risk there is of bacterial spoilage or oxidation. This is particularly important in a hot climate, when the process of deterioration is much faster than in cool regions. Insects, including Drosophilidae (vinegar flies), are immediately attracted to damaged fruit. Should there be any delay in processing, protecting the bins of fruit with dry ice will help prevent deterioration.

25.9 Selection and sorting

The purpose of sorting is to remove unripe fruit and altered fruit, that is fruit affected by rot. Sorting can take place at a number of points:

1 roguing the grapes while on the vine – particularly useful if they are to be mechanically harvested;
2 selection by the pickers, who may be instructed as to which fruit to harvest and which to leave on the vines;
3 sorting at a table on a trailer situated at the ends of the rows of vines, the fruit being sorted manually by people on each side of a moving belt;
4 upon arrival at the winery clusters or, in the case of machine harvesting, berries, sorted manually by people on each side of a moving belt or vibrating table;
5 after de-stemming, in the case of hand-picked grapes.

There are now several designs of mechanical sorters, including those that identify the differing masses of healthy and damaged grapes, and these can be particularly valuable where the costs or skills of labour are a major issue. Optical sorting machines are increasingly used, but their high cost may prohibit their availability to the small estate. A machine that works by separating grapes by grape density, the 'Tribaie', has proved very successful particularly with mid-sized producers, and there are now 120 of these in use around the world. Vibrating and moving belt tables remain a popular and

Figure 25.3 Hand sorting.

efficient alternative for producers of all sizes. A photo of hand sorting taking place in Argentina is shown in Figure 25.3, a vibrating sorting table in Figure 25.4 and the Amos Industrie Tribaie density sorter in Figure 25.5.

25.10 Use of pumps/gravity

Pumps are, of course, used extensively in wineries for tank filling, pump-overs (if practised), racking, bottling, etc. They are a physically aggressive way of moving liquids, and their use for moving must can be particularly detrimental, breaking seeds and causing an increase in dissolved oxygen. The latest designs of peristaltic pumps are much gentler than traditional impeller and centrifugal pumps. Figure 25.6 shows a range of impeller and centrifugal pumps, and Figure 25.7 shows a peristaltic pump. However, wineries should be designed or adapted to enable more gentle movement by gravity as far as possible. For instance, fruit from the crusher can be discharged into a small

Figure 25.4 Sorting at a vibrating sorting table.

Figure 25.5 Tribaie sorter. Source: Courtesy of AMOS INDUSTRIE.

Figure 25.6 A range of centrifugal and impeller pumps.

Figure 25.7 Peristaltic pump.

tank, which is then lifted by hoist to gravity-feed the fermentation vats or, in the case of white wine, the press. In the production of red wines, if the fermentation vats are high enough above the ground, at pressing time a horizontal press or the cage of a vertical press may be moved to the discharge hatch of each vat and filled, with the minimum of movement of the skins.

25.11 Control of fermentations and choice of fermentation vessel

It is important that temperature be controlled throughout the fermentation period. Although temperature control systems for large fermentation vats are now commonplace, challenges are still posed in adjusting the ferment to the desired temperature (normally cooling but occasionally warming) for particular stages of fermentation. Temperature is not uniform throughout the vat – heat rises, and regular equalising of the ferment is desirable. Barrel ferments, although having a greater percentage surface area for heat discharge, can present problems, too, and ambient cooling and humidity systems are highly desirable. We recall that white wines generally require fermentation under cool conditions (10–16°C), especially if an aromatic style is required; slightly warmer conditions (17–20°C) are required for full-bodied styles. However, for good colour and flavour extraction for reds, the winemaker may wish to ferment at 30–32°C, particularly during the early and middle fermentation periods. The tolerance range below the point at which there is a danger of 'stuck' fermentation is not very wide, and examples of cooked flavours from overheated ferments are all too common. Stuck ferments are much less of a problem than in the past when hot autumns posed particular challenges. There are many recorded instances of stuck fermentations at *cru classé* properties in Bordeaux before the advent of efficient temperature control systems, spearheaded in the 1960s at Château Haut Brion and Château Latour, but not commonplace elsewhere until the 1970s. There are still properties lacking temperature-control equipment in old cuboid vats made of cement!

As we saw in Chapter 8, fermentation vats can be made of stainless steel (inox), wood or concrete. Stainless steel was without doubt the material of choice for the majority of winemakers until a few years ago, but there is a move back to wood or concrete. Stainless steel is easy to clean and maintain, but care must be taken to avoid reductivity in some fermentations. Concrete vats, once mostly cuboid, may now be purpose designed to the winery's specification, and there is an endless number of possible shapes. Vats at Château Cheval Blanc, Saint-Émilion Grand Cru Classé (A) are shown in Figure 25.8, at Château Prieurie-Lichine, Margaux Grand Cru Classé in Figure 25.9 and Anfora (amphorae) vats at Famila Zuccardi in Valle de Uco, Argentina in Figure 25.10. Concrete has excellent thermic insulation properties and, if unlined (other than

Figure 25.8 Vats at Château Cheval Blanc.

Figure 25.9 Vats at Château Prieurie-Lichine.

Figure 25.10 Anforas at Zucccardi, Argentina. Source: Courtesy of Zuccardi, Argentina.

a light spray of cream of tartar), also allows minute oxygen permeation, and many winemakers who would micro-oxygenate when using stainless steel vats say that concrete eliminates the need for this. Concrete eggs keep the wine in circulation, and when lees-ageing white wines, the amount of contact is increased, impacting, inter alia, upon texture.

25.12 Use of gases

At many stages during the winemaking process, inert gases may be used to maintain freshness and prevent oxidation and other spoilage. The gases commonly used are nitrogen, carbon dioxide and, increasingly, argon. Carbon dioxide is highly soluble in wine, and its use can give the finished wine a 'prickle' on the tongue. This is particularly undesirable in red winemaking. Nitrogen, however, having a lower molecular weight does diffuse quickly. Accordingly, mixtures of two of the gases have advantages.

Horizontal tank presses and fermentation vats may be sparged with inert gas prior to filling with grapes, crushed grapes or fermented grapes. Partially filled vats may be blanketed by gas, and when racking and preparing for bottling, empty vessels may also be flushed with gas. Considerable amounts are required to flush all the air from the vats. The average quantity of gas used in producing a (9-litre) case of Australian wine is 10 litres. For high-quality wine, this rises to 600 litres.

25.13 Barrels

For those wines that benefit from oak ageing, there is no doubt that barrels are infinitely superior to the oak treatments discussed in Chapters 12 and 24. The impact of the barrels on the style and quality of wine will depend on a number of factors including size of the barrel; type and origin of oak (or other wood); manufacturing techniques including toasting; amount of time spent in barrel; and where the barrels are stored. There is also a marked difference between the maturations of the so-called château barrels, using fine-grained oak that has a thickness of 22 mm, against the standard 'export' barrel with a thickness of 27 mm. Most quality-conscious producers prefer to source their barrels from several coopers, as each has its 'house style', which will impart subtle differences to the maturing wine and give a greater range of components to influence the quality of the final blend. The oak from each individual forest has its own 'terroir' character, and a producer may specify the particular sources required to give the wine in each barrel the required characteristics. Figure 25.11 shows a barrique made from oak from Paris-Halatte at Château Lassegue in Saint-Émilion.

Figure 25.11 Barrique from Paris-Halatte at Château Lassegue, Saint-Émilion.

Table 25.2 Guideline prices of oak barrels

Size (litres)	Type	Type of oak	Price	Cost/litre[a]
225	Bordeaux Export	French	£560[b]	£2.48
225	Bordeaux Export	American	£420	£1.86
225	Château	French	£660	£2.93
300	Transport	French	£710	£2.36
500	Transport	French	£1010	£2.02
300	Transport	70% French, 30% Euro	£555	£1.85

[a]The cost-per-litre figure quoted is a simple calculation and writes off a new barrel after one use. Of course, barrels can be used again (second, third and maybe even fourth fill). However, after adding the handling and wine loss costs detailed herein, the cost of barrel ageing in new wood equates approximately to the cost per litre stated.
[b]Based on wood seasoned for 2 years; 3-year seasoned oak is more expensive.

The approximate costs of new 225-litre barriques in 2015 are detailed in Table 25.2. The cost of barrel ageing is not just that of the barrels but also that associated with loss of wine, labour for all the operations required and, of course, the finance involved in time and space. After a barrel is filled, the wood will absorb some 4 or 5 litres of wine. As a result of evaporation, the barrel requires regular topping up, although some oenologists question the effectiveness of this exercise if the barrels are sealed. Evaporation losses amount to 2–4% per annum, depending, inter alia, upon the temperature and humidity of the barrel store. Most winemakers undertake regular racking, perhaps six times for wines that spend 18–21 months in barrel, which will also result in small losses of wine. However, some producers have reduced the number of rackings, believing that minimal handling preserves integrity. Finally, there will be a loss of 1 or 2 litres when the barrels are emptied. Thus, during an 18-month period in barrel, the losses of wine will amount to 14 litres, or approximately 5% of the barrel volume.

25.14 Selection from vats or barrels

The production of top-quality wine involves making selections throughout the vinification process – from arrival of fruit at the winery to the components of the finished wine. The exclusion of vats or barrels containing wine from younger vines or from poorer parts of the vineyard, or that which simply has not turned out to be of a very high standard, is most desirable. The excluded wines can be sold as a 'second wine'. Alternatively, they may be moved down a level in a multi-tiered range of wines or, if substantially inferior, sold to a bulk producer as a blending component. It is in the selection

Table 25.3 Conditions needed for wine storage

Darkness	Apart from the fact that light is usually also a source of heat, exposed bottles can suffer from 'light-strike'. White wines will darken, and red wines become more brown.
Laying down	Bottles should be stored on their sides in order to keep the corks moist and thus expanded in the neck of the bottle. Wines with plastic or screw-cap closures should be kept standing up.
Constant temperature	Not always easy to achieve. A constant temperature promotes a controlled and balanced maturation, and corks that expand and contract with temperature changes are at risk of failure.
Temperature of 13°C (55°F)	This is often referred to as 'cellar temperature'. A couple of degrees' discrepancy either way will not pose a major problem. Wines mature quicker, and less evenly if the temperature is much above this. Constancy is more important than the actual temperature.
Away from strong smells	These can be absorbed through the closure

and blending process that the advice of winemaking consultants such as Michel Rolland proves valuable.

25.15 Storage

Whatever the quality of wine immediately after bottling, the way in which the bottles are subsequently stored will impact significantly on the quality at the point of opening. Some wines have a legal minimum period of ageing in bottle before being released for sale; e.g. Rioja Gran Reserva (red) must be bottle-aged for at least 2 years (with at least 5 years' total ageing including at least 2 years in cask). However, many producers wish to give their wines at least a short period of time in bottle before release from the cellars, even if not legally required to do so.

Storage conditions for wines laid down for bottle maturation need to be particularly suitable, but even wines designed for immediate drinking can be irreparably damaged if exposed to excesses of light, heat or cold. Details of suitable storage conditions are shown in Table 25.3.

CHAPTER 26

Selection by buyers

In this chapter, we will look at some of the considerations trade buyers may take into account when selecting wines for resale. We consider the important role of supermarkets in the overall wine market, and the renaissance of specialist wine suppliers. Also discussed are crucial price points and the decisions as to style and individuality that a trade buyer must make. Finally, we cover wine specifications and necessary analysis. Most of the market examples given are taken from the UK, but readers from around the world should have little difficulty in applying these to their own country. According to figures from OIV, the UK is the second largest importer of wine by value in the world, valued at €3.7 billion in 2013, being just behind the USA at €3.9 billion. However, Germany remains the largest importer by volume, taking some 15.4 million hectolitres (value €2.6 billion) compared with the UK's 13.1 million hectolitres. Put simply, Germany imports mostly inexpensive wine.

Historically, the marketing and sale of wine were the domain of wine merchants who would supply trade customers and/or the public from stocks sourced from long-established contacts. The wines from France and Germany dominated their lists, and 'lists' they were, for many wines, particularly the more expensive, would be hidden from the customer's sight. Most customers, not just the novice, would likely have felt intimidated when faced with the gentleman in the pin-striped suit who clearly demonstrated their vastly superior knowledge. Today, it is very different. While there are some excellent regional merchants, their strength lies in wines that cost £8 or more, and it is the supermarkets who dominate the volume market. Supermarkets de-snobbed wine, making buying easy and female friendly, and sales rocketed. However, producer/suppliers now lament that although the overall UK market

Wine Production and Quality, Second Edition. Keith Grainger and Hazel Tattersall.
© 2016 John Wiley & Sons, Ltd. Published 2016 by John Wiley & Sons, Ltd.

achieved decades of continuous and consistent growth until 2007, volume sales have declined by 14% since then, as detailed in Table 26.1. Table 26.2 shows the position in the UK sales 'league table' of the 10 most important countries of origin for the years 2013 and 2008.

Crucially, the number of important trade buyers of everyday wines has diminished. This number can almost be counted on the fingers of one hand, owing to mergers, acquisitions and the incessant rise in supermarket power and market share. However, in the last couple of years, supermarket growth, too, has stalled, ranges are being decimated and the number of suppliers is being severely pruned.

Table 26.1 UK sales of wine for the years 2007–2014

Year	Cases (×1000) of 9 litres (12 standard bottles or equivalent)
2014	140,365
2013	142,489
2012	144,290
2011	147,129
2010	151,474
2009	154,117
2008	151,295
2007	164,202

Extrapolated by the authors from figures quoted by The IWSR/Vinexpo/OIV.

Table 26.2 UK wine sales by country of origin for the years 2013 and 2008

Year 2013			Year 2008		
Position	Country	Cases (×1000) of 9 litres	Position	Country	Cases (×1000) of 9 litres
1	Australia	24,286	1	Australia	29,708
2	France	17,352	2	USA	15,785
3	Italy	17,200	3	France	14,192
4	USA	14,825	4	Italy	12,153
5	Spain	13,566	5	South Africa	9954
6	South Africa	11,091	6	Chile	7047
7	Chile	10,790	7	Spain	6493
8	New Zealand	4551	8	Germany	4005
9	Germany	3416	9	New Zealand	2102
10	Argentina	2376	10	Argentina	1267

26.1 Supermarket dominance

To the wine lover, the image is an endearing one. The supermarket buyer is travelling from one small wine estate to the next, sidestepping chickens in the yard and the obligatory vineyard dog, to taste the wines as the anxious host looks on. Suddenly comes the wide beam on the buyer's face, the 'eureka' moment, and he proudly informs the beaming owner: 'This is the one – I'll take all you have'.

In the real world, things are somewhat different. There is a joke among wine producers that a supermarket buyer is assessing one of the wines in particular detail. The nose goes in and out of the glass; the wine is swirled and tasted again, followed by that long reflection. There is some scratching of the head. Finally comes the pronouncement: 'Sorry – there's just not enough margin!' Whether or not a wine is listed will depend not just on the actual product and sometimes an audit of the producer's facilities, but also on listing allowances, pre-investment, promotional budgets, exposure funds, market support, supplements for gondola ends, retrospective discounts, etc.

In Germany, Aldi and Lidl sell approximately one bottle in two of all wines in the country. In the UK back in 2008, according to *Nielsen*, supermarkets accounted for 68% of total wine sales by value and over 70% by volume. By 2014, the figures had fallen to 62% by volume and 60% by value. However, such figures are subject to challenge, as *Nielsen* only records electronic point-of-sale data. This perhaps gives a very distorted picture of the UK market, as they include almost all supermarket sales but very little of the sales of specialist independent and online wine merchants, leading one to believe that:

1 there is little market for fine wine;
2 everyone gets most of their wine from supermarkets;
3 the average price per bottle is very low;
4 big-volume brands dominate the market.

'The public wants what the public gets' is a line from a 1970s popular anthem by the band Jam, and this has never been more true, particularly in so far as the unimaginative range of many supermarkets is concerned. Twenty years ago, the real diversity of wines offered was much greater than today, even though now the supermarkets boast of having more labels. Let us take south-west France as an example. At that time, the UK shopper could easily find some eclectic and distinctive wines, for instance: Jurançon, sweet spicy and tropical, made from the Petit and Gros Manseng grape varieties; white Gaillac, somewhat nervy, made from a blend of mostly local varieties including Len de l'El, Mauzac and Ondenc; Madiran with the wonderful leathery anis flavours of the predominant Tannat variety; Irouléguy, deep, dark and tannic and again made mostly from Tannat; and many more. Today, their place has been taken on the shelves by ubiquitous South-Eastern Australian Chardonnay and Cabernet Shiraz, perhaps bearing many different labels including the supermarkets' own, but in reality the bottles contain similar wine, as they are produced by one or other of the 'super' wine factories. The reds are fruity and soft. The whites are fruity and soft. Classic regions, too, have lost listings; for example, the number of wines from Bordeaux listed by supermarkets is very small compared with the production (approximately 800 million bottles per annum) and the worldwide importance of the region.

Of course, supermarkets do list many good-quality wines that show regional identity, but buyers often have safety in mind. They avoid wines that have such misunderstandable characteristics that might result in returns, as witnessed by a buyer's aide-memoire on the wall of a supermarket's tasting room: 'reject wines with sediments' and 'reject garnet or tawny reds', even though such wines could be classic examples. It should be noted that some supermarkets do have a particularly fine, if small, 'inner cellar selection' of top-quality wines available in flagship stores.

The largest mail-order retailer in the UK is Laithwaites, which trades under many names including many wine 'clubs', e.g. British Airways Wine Club. Laithwaites now trades globally, and turnover exceeds £350 million. The Wine Society, which is a members' cooperative, has over 123,000 active members, and in the fiscal year ending January 2015, sales of some £90.5 million, with members spending an average of £730 per annum, including 'opening offers' of unfinished wines released for sale by properties. Online sales are now significant for both of the above retailers, defying the general worldwide picture – Internet sales are still a mere 5% of the global wine market.

During the last decade, there has been a re-emergence of independent suppliers, whose passion for the product is as much, if not more, a driving force than the incessant drive for maximum profit and shareholder value. Not answerable to city institutions, and in some cases with the emergence of crowdfunding, they are less dependent upon banks and conventional sources of finance. They have the freedom to buy eclectic wines on the basis of quality and individuality, and some choose to invest in their producer suppliers. Naked Wines, recently purchased by Majestic but funded by 250,000 'angels', saw turnover increase by 40% in 2014 and, at the time of writing, this stands at £70 million.

26.2 Price point/margin

There are many crucial price points for wines, and producers and retailers break through them at their peril. The selling price point and profit margin are therefore major considerations when selecting potential listings. It is a sad, even depressing, fact that at the time of writing, according to figures extrapolated from IRI and Nielsen (and the reader should note the caveat above), the average selling price for a 75 cl bottle of wine in a UK supermarket is £5.41. Multiple specialist merchants sell at an average price of £5.60 per bottle, but taking the off-sales market as a whole, that is sales other than in pubs, wine bars, clubs and restaurants, the average price is still only £5.46 (again revealing the dominance of supermarkets). It is interesting to note that the country with the highest average price (£7.37) per bottle is New Zealand, relatively lowly placed at number 8 in the UK sales league table, but with high consumer perception of quality, particularly of wines made from Sauvignon Blanc and Pinot Noir. With regular rises in excise duty, which retailers expect suppliers to fund, the pressure on producer margins is greater than ever. Retailer margins are healthy: it is not possible to quote an average figure, but many retailers achieve 30–40% profit on return – indeed they can afford to offer across-the-board discounts of 25% on certain weeks of the year. Nieslen and IWSR figures state that 60% of wines sold in the UK are at a 'promotional' price, although it must be said that many of the 'list' prices are hugely inflated in order to allow for such promotions. Supermarkets and many of their customers have become locked in a promotional cycle, with consumers becoming bargain junkies, knowing the promotional price of everything but the value of nothing. The on-trade, of course, looks for much greater margins, and the published accounts of one large group of public houses and gastro-pubs (which offers a wide selection of wines by the glass) reveal that they are working on 55% profit on return.

26.3 Selecting wines for market and customer base

Independent wine merchants have often been heard to say when choosing an outstanding but obscure wine for their list: 'If I can't sell it, I'll be happy to drink it'. Partly owing to the Internet giving the independents a national rather than restricted local market, the number of independents is again increasing after years of decline. However, for many independent wine merchants, knowing their customers, understanding their aspirations and selecting wines that fit with these are the keys to success. Multiple specialists, too, have researched their customer profiles and even brand outlets to suit. Wines are chosen to reflect customers' perceived needs and spending patterns. Restaurant lists, often written by their wholesale supplier, need to have regard for the client base, the profile and, of course, the menu.

26.4 Styles and individuality

Any wine list should exhibit a broad range of styles and prices. At low price points, or 'entry level wines' to use the marketeer's parlance, the concept of individuality is non-existent, but the economies of scale make possible clean and fruity wines for those to whom wine is little more than a commodity. For wines made from a single variety, the character should be apparent, and dual-variety blends should express both components as well as constitute a harmonious whole. It is true to say that overproduction of grapes, as is presently the case in many countries of both the New World and Europe, has resulted in a plentiful supply of sound fruit for such wines. Even if the label states a district of origin, this does not guarantee that the wine will show the identity of the area. It is unrealistic to expect that, for example, a Minervois retailing at little more than £5 per bottle will be anything other than a simple drink, but with a hint of the character of the Languedoc-Roussillon region. However, even at relatively low price points, fraudulent labelling does still exist. Historically, it was the illustrious appellations whose wines commanded high prices that were often compromised, but now this is much less the case. Today, it is fashionable varieties such as Pinot Grigio (at present the best-selling variety in the UK) from popular regions such as Veneto that are likely to be the subject of 'incorrect' labelling. Far more Italian 'Pinot Grigio' is sold than the country produces!

As we start to progress up the price scale (and we should perhaps bear in mind that most consumers never leave the so-called 'entry level'), there is a broadening of available styles and the possibility of regional, district and producer identity. The price at which real regional characteristics become apparent depends very much on the region in question. For example, it is possible to find Vouvray retailing for around £8 that shows true Chenin Blanc peach character, lively and racy with the classic mineral tones of this part of Touraine. A true-to-type Pouilly Fumé, produced further up the Loire in the Central Vineyards area, should be flinty, gently smoky, crisp and elegant. It will have a restrained Sauvignon Blanc nettle and gooseberry character but will have a starting price of £13 or more. From around this price point, wines should express true individuality, they should excite and stimulate, and they should say who they are, why they are and precisely where they are from.

26.5 Continuity

Many consumers today expect products to be continuously available, and indeed the very concept of seasonality, the eager anticipation of the new season's lamb, cherries or peas hardly exists. This applies to wines, too, and the excitement of drinking Beaujolais Nouveau in November from grapes that had still been on the vine in September is long gone. For the retailer or restaurant, a lack of continuity of a product can present a major headache, with additional costs in changing lists, etc.

Table 26.3 Top 10 wine-producing countries and production figures (all figures in thousands of hectolitres)

Country	2014	2013	2012	2011	2010
France	46,698	42,004	41,548	50,757	44,381
Italy	44,739	52,029	45,616	42,772	48,525
Spain	41,620	45,650	31,123	33,397	35,353
USA	22,300	23,590	21,650	19,140	20,887
Argentina	15,197	14,984	11,778	15,473	16,250
Australia	12,000	12,500	12,260	11,180	11,420
South Africa	11,316	10,982	10,569	9725	9327
China	11,178	11,780	13,511	13,200	13,000
Chile	10,500	12,820	12,554	10,464	8844
Germany	9344	8409	9012	9132	6906

Information obtained from OIV reports.

Historically, wholesalers would hold considerable stocks of both young and mature wines, and could, to a large degree, balance supply to their customers. Today, producers increasingly supply larger trade customers directly or use the services of an agent or sales office that holds no stocks. But modern customers require continuity of supply and stability of price. This may be planned, and financed, within the available stocks of a vintage, but it is unrealistic to expect such guaranteed continuity from year to year. Wine is an agricultural product and as such is at the mercy of the weather, diseases and other natural forces that can result in considerable annual variation in both the quantity and quality. This is particularly true in the Old World. For example, the crop in Italy in 2014 was 14% down on the 2013 figure, resulting in Italy losing its position as the world's largest wine producer for that particular year. Table 26.3 shows the quantity of wine made in the 'Top 10' producing countries in the years 2008–2014.

However, New World countries are not immune from huge crop variations. For example, the 2007 total wine grape harvest in Australia, following a growing season affected by continuous drought, frosts and bushfires resulting in smoke taint, was 1.39 million tonnes. This figure was down from 1.9 million tonnes in 2006, which equates to a drop in the amount of wine produced from 1410 million litres in 2006 to 955 million litres in 2007, a reduction of over 30%. The 2013 figure was 1.75 million tonnes. It is obvious that the dangers as far as continuity of supply to a market are very real. Large producers often hold surplus stocks of commodity wines in tank from previous vintages, for release to the market when needed to help counter annual production variations. However, the impact upon the amount of product available to meet key price points cannot be understated.

26.6 The place of individual wines in the range

When selecting wines for inclusion in a range, the buyer needs to have in mind the other wines already listed and, unless existing wines are to be directly replaced, how the new selections will sit against them. It is unlikely that the consumer will buy more wine simply as a result of extra listings, and it is probable that each wine added will take the sale from one already stocked. At the higher price points, each wine will need to have not only quality but also individuality and to earn its place on a list by saying something that its peers do not.

26.7 Exclusivity

It is easy for any retailer or restaurant to build a list around large and well-known brands. Such lists might be comforting but lack identity and interest and will not have any competitive advantage, other than possibly if the seller can offer an attractive price. While large brand owners may finance promotions and publicity, this has to be counterbalanced against the inbuilt cost of these activities. Selecting individualistic wines that competitors do not have creates interest and can give a competitive edge and healthy margins, for the customer has no direct price comparisons. This can be very important for restaurants whose customers resent paying three times the price for a wine readily available on the supermarket shelves. Sourcing and retaining lines that are exclusive to the retailer (always staying within anti-competition legislation rules) can prove challenging but rewarding. Producers will often prepare an alternative label for those buying sufficient quantities, or the retailers may wish to have ownership of their own label. Buyers' Own Brands are now a very important means of guaranteeing exclusivity, and of course the wine may be prepared to the customer's specification. It is worth noting that producers, too, may wish to offer an alternative label as a means of disposing of excess stocks at a price lower that their usual label, without devaluing the status of the winery's own label. 'Cleanskin' wines, which are just labelled with the minimum legal information, and usually the grape variety, are particularly popular in Australia. The consumer believes they are getting product from top wineries at a bargain price, but in many cases the wine sold is of a quality that the winery might not wish to put its name to.

26.8 Specification

Producing wines to buyers' specifications may seem to be a very modern phenomenon, but this has been taking place, in a much less technical way, for centuries. Put simply, a producer would want to make the product as appealing as possible in the market in which it was to be sold. For example, the wines of Bordeaux have long been exported to Britain, a trade well established in the 300 years (1152–1453) that Bordeaux was English. However, customers often found the wines too light for their palate (the name claret used for the red wines of the region comes from a French word *clairet* meaning pale red). Accordingly, the tradition grew of blending in some deeper and fuller-bodied wine from the warmer Rhône Valley, a practice known as *hermitagé*, although it is unlikely that the additive wines came from the great vineyard of Hermitage. Also, before the wines were shipped in cask, they would have a little spirit from the neighbouring region of Cognac added to give them an alcoholic boost and to help stabilise them for their sea journey. Thus, the style of wine was adjusted for the market. Burgundy, too, was often adjusted for the *goût anglais*. The railway station at Beaune became an important link in the production chain, for here were received bulk quantities of cheap full-bodied wines from warmer regions further south, particularly the Rhône Valley, to be blended in with the often thin wines from vines that were over-yielded. Thus, the wines were adjusted to the specification of the buyer (full body) and the seller (reduced production cost).

Today, the supermarket and multiple specialists, especially when considering a buyer's own brand, may have a much more detailed specification as to style and aspects of technical composition, particularly regarding the levels of alcohol, acidity and residual sugar. With modern production techniques in the large wineries, fulfilling the requirements presents few challenges. For example, the demand for deep-coloured, red wines with soft tannins at low price points has coincided with technology that makes these achievable, including flash détente and thermo détente. Containers and closures are specified, too – in the UK, several supermarkets specify that their producer/suppliers use screw-caps or other synthetic closures.

26.9 Technical analysis

Until recently, there were many thousands of small producers throughout the world who never submitted their wines to more than a basic analysis. The wines from some of these were truly excellent, and from others distinctly mediocre. Perhaps the only measurement equipment used during the wine-making process comprised weighing scales, capacity measures, a thermometer and some hydrometers. By differential calculation of the specific gravities, that is of the start must weight and that of the wine after fermentation, the percentage alcohol of the finished wine was determined to the accuracy needed for labelling.

All wines now require at least a basic laboratory analysis to ensure that they comply with the legislation in the country of production and sale, including specific requirements for PDO wines if applicable. Interestingly, EU regulations require that PDO wines be subject to tests for 'wine *behaviour*' (authors' italics). At the time of writing, the only 'technical' information required on the label of a wine sold in the UK is the content and the percentage alcohol to an accuracy of 0.5%. Additionally, since November 2005, labels must state allergen information that the wine contains sulfur dioxide (sulfites), and since 1 July 2012 under Regulation (EU) No. 1169/2011 [since amended by (EU) 78/2014] egg and milk products (commonly used as fining agents). Generally, ingredient labelling is not required. From a marketing point of view, this may be a good thing for much as perceived as a natural product, EU regulations detail 50 permitted additives (the permitted list in USA totals 65). There are, of course, legal limits specified in EU regulations on the amount of the many chemicals that may be present in or added to a wine. It should be noted that further restriction exists in some member states. Third countries have their own regulations, and without detailed analysis producers are

Table 26.4 Parameters used in the analysis of wine

2,3,4,6-Tetrachloroanisole (ng/l)	Haze (formazine turbidity unit)
2,4,6-Tribromoanisole (ng/l)	Hydrogen sulfide (µg/l)
2,4,6-Trichloroanisole (ng/l)	Iron (mg/l)[a]
4-Thylphenol (µg/l)	Lactic acid (g/l)
Acetaldehyde (mg/l)	Lead (mg/l)[a]
Alcohol (% vol. at 20°C)[a]	Magnesium (mg/l)
Arsenic (mg/l)[a]	Malic acid (g/l)
Ascorbic acid (mg/l)[a]	Ochratoxin A (µg/l)
Bacteria [colony-forming units (CFU)]	Optical density (420 and 520 nm)
Benzoic acid (mg/l)	Pentachloroanisole (µg/l)
Bitterness (bittering units)	pH[a]
Brettanomyces spp.	Potassium (mg/l)[a]
Calcium (mg/l)[a]	Protein stability[a]
Calorific value (kJ/100 ml; kcal/100 ml)	Reducing sugar (g/l)
Carbon dioxide (g/l)	Silver (mg/l)
Cold stability[a]	Sodium (mg/l)[a]
Contents (ml)	Sorbic acid (mg/l)[a]
Copper (mg/l)[a]	Specific gravity[a]
Density (g/l)	Sugar-free dry extract (g/l)[a]
Discolouration	Total (titratable) acid (g/l)[a]
Dissolved oxygen (mg/l)	Total dry extract (g/l)
Ethyl carbamate	Total phenolics (mg/l)
Filterability	Total potential alcohol (% vol.)
Free sulfur dioxide (mg/l)[a]	Total residual sugar (g/l)[a]
Fructose (g/l)	Total sulfur dioxide (mg/l)[a]
Glucose (g/l)	Volatile acidity [g/l (acetic acid)][a]
Glycerol (g/l)	Yeast (CFU)

[a] A certificate showing analysis results is commonly required by British supermarkets.

exposed to the risk of non-compliance. Some of the chemicals present may have been used in the winemaking process, some derive from the grapes or are manufactured by the wine itself, and others are contaminants. For example, Ochratoxin A (OTA) is produced by some species of fungi that infect many crops including cereals, coffee and grapes. The toxin can cause liver and kidney damage, and is a possible carcinogen. EU regulation (EC) No. 1881/2006 specifies a legal OTA limit in wine of 2 µg/l. The same regulation stipulates a maximum limit for lead in wine of 0.3 mg/l. EU regulation (EC) No. 401/2006 lays down sampling procedures for mycotoxins in foodstuffs which, of course, include wine.

Table 26.4 lists analytical parameters, including the tests commonly required by UK supermarkets. When comparing the results of wine analysis, it is important to ensure that identical expressions of the measurement are used. For example, total acidity is usually expressed as grams of tartaric acid per litre, but in France and some other countries, it is often expressed as grams of sulfuric acid per litre (to convert a tartaric acid figure to sulfuric, it is necessary to divide by 1.531).

Appendix

WSET® Diploma Systematic Approach to Tasting Wine

APPEARANCE		
Clarity/brightness		clear – hazy / bright – dull (faulty?)
Intensity		pale – medium – deep
Colour	white rosé red	lemon-green – lemon – gold – amber – brown pink – salmon – orange – onion skin purple – ruby – garnet – tawny – brown
Other observations		e.g. legs/tears, deposit, petillance, bubbles
NOSE		
Condition		clean – unclean (faulty?)
Intensity		light – medium(–) – medium – medium(+) – pronounced
Aroma characteristics		e.g. fruits, flowers, spices, vegetables, oak aromas, other
Development		youthful – developing – fully developed – tired/past its best
PALATE		
Sweetness		dry – off-dry – medium-dry – medium-sweet – sweet – luscious
Acidity		low – medium(–) – medium – medium(+) – high
Tannin	*level nature*	low – medium(–) – medium – medium(+) – high
		e.g. ripe/soft vs unripe/green/stalky, coarse vs fine-grained

Continued

Wine Production and Quality, Second Edition. Keith Grainger and Hazel Tattersall.
© 2016 John Wiley & Sons, Ltd. Published 2016 by John Wiley & Sons, Ltd.

Alcohol	low – medium(–) – medium – medium(+) – high **fortified wines:** low – medium – high
Body	light – medium(–) – medium – medium(+) – full
Flavour intensity	light – medium(–) – medium – medium(+) – pronounced
Flavour characteristics	*e.g.* fruits, flowers, spices, vegetables, oak flavours, other
Other observations	*e.g.* texture, balance, other **sparkling wines (mousse):** delicate – creamy – aggressive
Finish	short – medium(–) – medium – medium(+) – long

CONCLUSIONS
ASSESSMENT OF QUALITY

Quality level	faulty – poor – acceptable – good – very good – outstanding
Reasons for assessment	*e.g.* structure, balance, concentration, complexity, length, typicity

ASSESSMENT OF READINESS FOR DRINKING / POTENTIAL FOR AGEING

Level of readiness for drinking / potential for ageing	too young –	can drink now, but has potential –	drink now: not suitable for ageing for ageing or further ageing	– too old
Reasons for assessment	*e.g.* structure, balance, concentration, complexity, length, typicity			

THE WINE IN CONTEXT

Origins/variety/ theme	*for example:* location (country or region), grape variety or varieties, production methods, climatic influences
Price category	inexpensive – mid-priced – high-priced – premium – super-premium
Age in years	answer with a number not a range or a vintage

Notes for students:
For lines where the entries are separated by hyphens – students must select one and only one of these options.

For lines starting with "*e.g.*" **where the entries are separated with commas** – the list of options are examples of what students might wish to comment on. Students may not need to comment on each option for every wine.

WSET Diploma Wine-Lexicon: *supporting the WSET Diploma Systematic Approach to Tasting Wine®*

DESCRIBING FLAVOURS
   Be accurate: *think in terms of clusters*
   Be complete: *don't just rely on lists of descriptive words; aim to describe the quality and nature of the flavours too*

Primary aroma/flavour clusters: *the flavours of the grape*

Key questions		Descriptive words
	Floral	acacia, honeysuckle, chamomile, elderflower, geranium, blossom, rose, violet, iris
	Green fruit	green apple, red apple, gooseberry, pear, peardrop, custard apple, quince, grape
	Citrus fruit	grapefruit, lemon, lime, (juice or zest?), orange peel, lemon peel
Are the flavours delicate *or* aromatic? simple/neutral *or* complex? generic *or* well-defined? fresh *or* cooked/baked? under-ripe *or* ripe *or* over-ripe?	**Stone fruit**	peach, apricot, nectarine
	Tropical fruit	banana, lychee, mango, melon, passion fruit, pineapple
	Red fruit	redcurrant, cranberry, raspberry, strawberry, red cherry, red plum
	Black fruit	blackcurrant, blackberry, bramble, blueberry, black cherry, black plum
	Dried fruit	fig, prune, raisin, sultana, kirsch, preserved fruits
	Herbaceous	green bell pepper (capsicum), grass, tomato leaf, asparagus, blackcurrant leaf
	Herbal	eucalyptus, mint, medicinal, lavender, fennel, dill
	Pungent spice	black/white pepper, liquorice, juniper

Secondary aroma/flavour clusters: *the flavours of winemaking*

Key questions		Descriptive words
Are the flavours from yeast, MLF, oak *or* other?	**Yeast (lees, autolysis, flor)**	biscuit, bread, toast, pastry, brioche, bread dough, cheese, yoghurt
	MLF	butter, cheese, cream, yoghurt
	Oak	vanilla, cloves, nutmeg, coconut, butterscotch, toast, cedar, charred wood, smoke, resinous
	Other	smoke, coffee, flint, wet stones, wet wool, rubber

Continued

Tertiary aroma/flavour clusters: *the flavours of time*

Key questions	Descriptive words	
	Deliberate oxidation	almond, marzipan, coconut, hazelnut, walnut, chocolate, coffee, toffee, caramel
Do the flavours show deliberate oxidation, fruit development *or* bottle age?	**Fruit development (white)**	dried apricot, marmalade, dried apple, dried banana, *etc.*
	Fruit development (red)	fig, prune, tar, cooked blackberry, cooked black cherry, cooked strawberry, *etc.*
	Bottle age (white)	petrol, kerosene, cinnamon, ginger, nutmeg, toast, nutty, cereal, mushroom, hay, honey
	Bottle age (red)	leather, forest floor, earth, mushroom, game, cedar, tobacco, vegetal, wet leaves, savoury, meaty, farmyard

Other observations: *sweetness, acid, tannin, alcohol and texture*

	Sweetness	austere, thin, drying, unctuous, cloying, sticky
Use sparingly to create a more complete description do not use instead of low – medium – high *etc.*	**Acidity**	tart, green, sour, refreshing, zesty, flabby
	Alcohol	delicate, light, thin, warm, hot, spirity, burning
	Tannin	ripe, soft, unripe, green, stalky, coarse, chalky, grippy, fine-grained, silky
	Texture	stony, steely, mineral, oily, creamy, mouthcoating

CONCLUSIONS **ASSESSMENT OF QUALITY** Use evidence: *don't just give an opinion, every comment must be backed up with evidence* Be comprehensive: *comment on all of the key elements that contribute to quality*	**Note to students: *in certain papers the examiners ask for other concluding remarks too. Read the Candidate Assessment Guide for details.***

Key questions	Key elements		
How well balanced are the components of the wine?	**Structural balance**	acid, alcohol, tannins	flavour, sugar
	Other	• intensity, length of finish • expressiveness	• complexity, purity • potential to age

Note to students: the WSET Level 4 Wine-Lexicon is designed to be a prompt and a guide for students. It does not attempt to be comprehensive and it does not need to be memorized or slavishly adhered to.

Glossary

Acetaldehyde   Product of the oxidation of ethyl alcohol; at low concentrations enhancing wine aromas, but at high concentrations giving a 'Sherry-like smell'.

Acetobacter   Genus of bacteria of the family Bacteriaceae. The organisms oxidise alcohol to acetic acid.

Acids   Essential components of grapes and wine, giving freshness and bite. Tartaric and malic acid comprise 69–92% of the organic acid of grapes. Citric, lactic, succinic and acetic acids are also present. Malic acid is naturally converted to the softer (yogurty) lactic acid via malolactic fermentation, which naturally tends to follow the alcoholic fermentation.

Agrafe   Lit. a staple. This metal clip was used in Champagne making to fasten and keep the cork in place during the second fermentation. It was used before the onset of the crown cap.

Alcohol   The product of fermentation, created by the enzymes in yeasts working on sugar. Alcohol is also added in neutral form during or after fermentation to produce fortified wines. Usually measured as percentage by volume [alcohol by volume (abv)]. See **Ethanol**.

Alcohol by volume (abv)   The measure of the alcohol (ethanol) content contained in a given volume of alcoholic beverage and expressed as volume per cent.

Anthocyanins   A group of polyphenols, found in the skins of red grapes, (and many other fruits and flowers) that is mainly responsible for the colour of red wines.

Appellation Contrôlée (AC)   Legal status within France's grading system [under European Union Protected Designation of Origin (PDO) wine regulations] that guarantees geographical origin and specifies production criteria for designated wines. See also **Appellation d'Origine Protégée (AOP)**.

Appellation d'Origine Protégée (AOP)   The highest French wine classification, from 2012 progressively replacing on labels, the previous AOC (AC) classification.

Assemblage   French term for the blending of different vats and parcels of wine to make up the final blend.

Wine Production and Quality, Second Edition. Keith Grainger and Hazel Tattersall.
© 2016 John Wiley & Sons, Ltd. Published 2016 by John Wiley & Sons, Ltd.

Astringent   A tactile sensation, mostly when tasting red wines that can be described as dry, sour, mouth puckering. Phenolics found in the skins and seeds of grapes are the cause of the sensation.

Auslese   Lit. 'selected harvest'. A German quality wine category for late-harvest wine and a riper category than Spätlese q.v.

Autolysis   Interaction between wine and solid yeast matter resulting in a distinctive bread- or biscuit-like flavour. Autolytic character is encouraged by ageing a wine on the lees, as takes place in the bottle during producer maturation of Champagnes, or in vat or barrel for certain other white wines. See also **Yeast autolysis**.

Balling   See **Brix**.

Barrel-ageing   Maturation in barrel in order to add oak, vanilla and toast flavours, and allow a controlled oxygenation.

Barrique   Small oak barrel holding 225 litres (300 × 75 cl bottles) of wine, traditionally used in Bordeaux; now increasingly used in other regions worldwide.

Bâtonnage   French term for the stirring of the lees in barrels during wine-making to improve texture and enhance flavours.

Baumé   Scale, commonly used in France, for measuring the density, being mainly the concentration of grape sugars, indicating the potential alcohol by volume (abv).

Beerenauslesen   A category of German wine made from a selection of individual overripe or botrytised berries. The grapes are so sugar-rich that it is impossible to ferment to dryness, resulting in wines that are naturally sweet. It is only possible to make the wines in very good years. Trockenbeerenauslesen is made from grapes that have become dehydrated, like raisins.

Bentonite   A type of clay that can be used for fining q.v. Bentonite has many other uses including as a lubricant for drilling bits in the oil industry, for tanking of ponds and as an enema.

Bidoule   A small plastic cup attached to the crown cap and placed with its open end facing into a Champagne bottle, after liqueur de triage has been added. It helps retain sediment to assist in the dégorgement process.

Blanc de Blancs   Champagne made exclusively from white grapes, likely to be 100% Chardonnay. The term is sometimes used by other sparkling wine producers but may include different white grape varieties.

Blanc de Noirs   Champagne made exclusively from red grapes: Pinot Noir and/or Pinot Meunier. This style of Champagne is always white. Some New World producers use this term for pink sparkling wines.

Bodega   A Spanish term referring to a winery, wine cellar or storehouse for maturing wine.

Botrytis cinerea   Fungus that can attack grapes in the form of destructive grey rot, or in certain climatic and weather conditions as beneficial 'noble rot', desirable for the production of sweet wines such as Sauternes and Beerenauslesen q.v.

Brix   Scale for measuring density that is mainly the concentration of grape sugars. Commonly used in the USA. 1 degree Brix is 1 g of sucrose in 100 g of the solution as percentage by mass.

Brut Nature   A Champagne bottled without dosage.

Butt   American oak (*Quercus alba*) 550-litre cask used for maturation of wines in the solera system during the production of Sherry.

Cambium   A layer of cells in plants that can produce new cells, including phloem q.v.

Cane   Mature shoot of a vine that has overwintered and turned orange-brown. Canes are cut back in winter according to the system of training used.

Canopy   Part of the vine that is above ground, including shoots, leaves and fruit.

Cap   The floating skins in a red wine undergoing fermentation.

Capsule   Once lead, now foil or plastic-film, which covers and protects the cork and bottle neck. Wax is also sometimes used, particularly for Vintage Port.

Carbon dioxide (CO_2)   By-product of fermentation. CO_2 is trapped in wine as bubbles by sparkling winemakers using the traditional, transfer or Charmat methods, and injected into very inexpensive sparkling wines.

Carbonic maceration   Fermentation method using whole bunches of black grapes. Under a blanket of CO_2, enzymic changes within the individual berries take place, resulting in an initial fermentation unrelated to yeast. Following this, the conversion of sugars by yeasts continues. Many light-bodied, fruity red wines are produced in this way.

Cask   Wooden container (usually cylindrical) used for fermenting or maturing wine and made in varying sizes. Usually made from various types of oak, according to the tradition of the region.

Centrifuge   Machine used to separate must or wine from the lees or other solids after fermentation.

Chaptalisation   The addition of sucrose or concentrated grape juice to the must or the juice in the early stages of fermentation to increase the alcohol level. In cool climates, grapes sometimes do not contain enough sugars to produce a balanced wine. The process is named after Dr Jean Antoine Chaptal, who was French minister of the Interior at the time of Napoleon.

Charmat   Sparkling wine production method, used for Prosecco and many inexpensive wines, in which the second fermentation takes place in tank. Named after its inventor, Eugène Charmat.

Chlorosis   Vine disorder resulting from iron deficiency in soil, often found in lime-rich soils. It can cause loss of colour in leaves, which turn yellow. Consequently, yields may fall as a result of reduced photosynthesis.

Clone   Group of organisms produced by asexual means from a single parent to which they are genetically identical.

CODEX An international directory compiled by OIV (q.v.) detailing the chemical organic and gas products permitted for the making and storage of wines.

Cold Soak Pre-fermentation cold maceration that involves aqueous extraction as opposed to alcoholic extraction of compounds from fresh pulp, skins and seeds into the must. Better colour, increased aromatics and fruitier wines are the aim.

Colloids Ultramicroscopic particles that do not settle out and are not removed by filtration, and so may cause haze in wine.

Colony-forming unit (CFU) A measurement, normally per millilitre (in the case of liquid food products) of the number of viable cells of microorganisms (bacteria, yeast and fungi). NB: A single living cell can reproduce by means of binary fission.

Concentrated grape must Grape juice that has been reduced to 20% of its volume by heating. If 'rectified', it has also had its acidity neutralised. An alternative to using sugar for chaptalisation q.v.

Concrete egg An egg-shaped concrete vat design, modelled on the ancient amphorae and increasingly being used in wineries. First commissioned in 2001 by Michel Chapoutier of the Rhône.

Congeners The colouring and flavouring matter in wines.

Consejo Regulador A locally controlled board in Spain under national government authorisation, responsible for classification and implementation of the designation of origin. Local growers and winemakers are represented on the board.

Cordon Permanent horizontal arm that extends from the trunk of the vine.

Coulure Poor setting of the flowers on a vine, usually owing to adverse weather conditions at flowering time (see also Millerandage).

Crossing Variety produced by fertilising one *Vitis vinifera* variety with the pollen of another *Vitis vinifera* variety.

Cru A French term for certain vineyards or communes regarded as superior.

Cru Bourgeois A classification of red wines produced in the Médoc and Haut Médoc districts of Bordeaux. The classification is revised annually.

Cru Classé An officially classified Bordeaux wine-producing property. The most famous classification is that of 1855, which classified the red wines of the Médoc district (together with Château Haut Brion situated in the Graves district) and the sweet white wines of Sauternes.

Crushing Process of breaking open the grapes ready for fermentation.

Cryoextraction A process where ripe grapes are picked, cleaned and refrigerated until frozen. The frozen grapes are then pressed, yielding a chilled must with high, concentrated levels of flavour.

Cuvaison Period of time a red wine spends in contact with its skins.

Cuve Commonly used French term for a vat or tank.

Cuve close Alternative name for Charmat or tank method used in sparkling winemaking.

Cuvée The word has many different meanings. It can mean the juice from the first pressing for Champagne or a blend and its components of any wine.

Débourbage Settling of solids (bourbes) in must or wine either pre- or post-fermentation.

Dégorgement Removal of yeasty sediment from the bottle towards the end of the traditional method of making sparkling wine.

Degree days System of climatic classification of viticultural areas developed by Amerine and Winkler of the University of California at Davis.

Délestage Translated as 'rack and return', a cellar operation in which a vat is completely drained of red wine, and then the fermenting wine is returned over the cap of skins. It is considered by many to be a gentler process than pumping over, giving softer, polymerised tannins and deeper colours.

Density of planting Number of vines planted in a given area of land, usually in Europe expressed as the number of vines per hectare.

Diurnal Lit. 'during the day'. Diurnal temperature range describes the variation in temperature, high to low, during a day. It generally increases with distance from the sea.

Dosage Following disgorging, the topping up of the bottles of sparkling wine produced by the traditional method q.v. with sweetened wine, according to the required style.

Dry extract The total of the non-volatile solids in a wine.

Dryland farming A method of farming practised in semi-arid areas without using irrigation. Drought-resistant rootstocks are usually necessary.

Egrappoir Machine that removes stalks from grapes, following which they may be crushed.

Eiswein German type of dessert wine produced from grapes frozen while still on the vine. The water content freezes (but not the sugars or dissolved solids). This allows a more concentrated must to be pressed from the frozen grapes, resulting in a very sweet and high-acid wine.

Élevage The 'rearing' or maturing of wines before bottling.

Esters A compound resulting from a reaction between acid and alcohol. Their presence in wine can give a smell of pear drops or bananas.

Ethanol Alcohol in wine and other alcoholic drinks, also known as ethyl alcohol. May be expressed as C_2H_5OH or CH_3CH_2OH.

Fermentation The conversion of sugars into alcohol through the action of yeasts.

Fermentation lock Device fitted to the top of a tank or barrel to allow escape of CO_2 but prevent ingress of air.

Filtration Passing the wine through a medium to remove bacteria and solids. A process that should be undertaken with care so as not to remove flavours.

Fining Removal of microscopic troublesome matter (colloids) from the wine before bottling. Materials that may be used for fining include bentonite, albumen (egg whites), gelatine and isinglass.

Flash détente   A technique whereby must is heated to 85°C and sent to a high-pressure vacuum chamber where the liquid is vaporised. The grape skins are deconstructed and the chamber cooled rapidly to 32°C. The process gives well-coloured, fruity red wines with soft tannins

Flavones   Colourless crystalline tricyclic compounds, the derivatives of which are plant pigments.

Formazine turbidity unit   The ISO adopted measurement of fine particles in a liquid by the method of light scattering.

Fortification   The addition of alcohol to certain wines (e.g. Sherry and Port) before, during or after fermentation.

Free-run juice   Juice naturally drained from solids, from a crusher, press or vat.

Fructose   One of two primary sugars contained in the pulp of wine grapes (glucose being the other). As the fruit ripens, levels of fructose increase, and it has a sweeter taste than glucose, thus being important in the making of dessert wines.

Glucose   One of two primary sugars contained in pulp of wine grapes (fructose being the other). It has a less sweet taste than fructose.

Goût de terroir   Lit. taste of the earth. A description used to describe a wine that shows the character of the environment in which it is produced.

Grand Cru   Lit. great growth. A French term for a wine-producing property or vineyard officially rated as superior. In Burgundy, however, the term refers to the finest vineyards.

Grand vin   General term for the 'top' wine from a producer, implying that it is the result of selecting the best component wines.

Green harvesting   Crop thinning that takes place after veraison but some time before normal harvesting to encourage the development and ripening of the remaining crop.

Gyroplatte   A mechanised riddling pallet used in sparkling wine production during the 'rémuage' process q.v.

Hectare   A measure of land, equalling 10,000 m² or 2.47 acres.

Hybrid   Result of the crossing of two different *Vitis* species, e.g. *Vitis vinifera* and *Vitis labrusca*.

Isinglass   Fining agent derived from fish.

Kabinett   A category of German wine, made from ripe but not overripe grapes, without chaptalisation q.v., usually light and delicate in style.

Kieselguhr   A coarse-grade diatomaceous earth powder used as a filtration adjuvant. Kieselguhr filtration will remove gross solids including lees.

Lateral shoot   Side shoots that develop from a bud (on a green shoot) and so taking nutrients from the vine. These are usually removed.

Lees   Sediment, including dead yeast cells, which settles at the bottom of the vat or barrel following the end of the fermentation process. Gross lees are coarse sediments, but lighter, fine lees may settle after initial racking.

Liège   Traditional cork closure used in bottling of Champagne.

Liqueur d'expedition   Liquid containing wine and cane sugar, or rectified concentrated grape must, used to top up bottle-fermented sparkling wine after dégorgement q.v.

Liqueur de tirage   Mixture of sugar (or rectified concentrated grape must) and yeast added to base wine at the time of bottling to induce secondary fermentation in traditional method sparkling wine production.

Maceration   The soaking of grape solids in their juice. This can take the form of cold soaking pre-fermentation or, in the case of red wines, post-fermentation. Maceration will help extract flavours and, for red wines, colour and tannins.

Malolactic fermentation (MLF)   A fermentation that may take place after (or occasionally during) the alcoholic fermentation, by which bacteria convert harsh malic acid into soft lactic acid. Almost without exception, it is desirable in red wines, but in white wines the winemaker may encourage or block it, depending upon the style required.

Marc   Skins, stalks and pips left after pressing grapes. May be distilled into brandy

Master of Wine (MW)   A member of the Institute of Masters of Wine who has passed the rigorous theoretical and tasting examinations and has had a dissertation accepted. At the time of writing, there are some 340 Masters of Wine worldwide.

Mesoclimate   The climate of a particular vineyard site.

Méthode champenoise   Now-defunct term replaced by 'traditional method'.

Mercaptans   Group of volatile, foul-smelling chemical compounds that can occur in wine. A symptom of reductivity q.v.

Mesoclimate   The climate of a particular vineyard site.

Microclimate   The climate within the canopy of leaves surrounding the vine.

Millerandage   Poor setting of fruit, resulting from poor pollination; consequently the grape berry fails to develop. See also **Coulure**.

Must   Unfermented grape juice, seeds, skins and pulp.

Must adjustment   The addition of various substances before fermentation to ensure the desired chemical balance. For example, this may include additions of tartaric acid (acidification), often practised in hot climates. De-acidification may be required in cool climates.

Must concentrators   Machines used to remove water from the juice of grapes. They can prove valuable following a wet vintage.

Must enrichment   Process before or in the early stages of fermentation whereby the sugar content of the must is increased to raise the alcohol level of the wine. It may be undertaken in cool climates where grapes struggle to ripen. This is strictly controlled by law in the European Union. See also **Chaptalisation**.

Must weight Measurement of density of must. In practice, this indicates the sugar levels contained in crushed grapes or juice. See also **Baumé**, **Brix**, **Oechsle**.

Mutage The addition of alcohol to stop fermentation – used in the making of sweet fortified wines.

Nanogram One billionth of a gram.

Node A point on the vine's stem at which a leaf, bud or branching shoot originates.

Non dosé A Champagne bottled without any dosage.

Noble rot *Botrytis cinerea* (q.v.) in its beneficial form.

Oak Preferred type of wood in which to ferment and/or mature wine. Provides character and imparts flavour. Barrels, staves or chips may be used according to winemakers aims and financial constraints.

Oechsle Scale used for measuring the specific gravity or must weight of grape juice to measure the final alcohol level of wine, commonly used in Germany.

Oenology (also Enology) The science of winemaking.

Oidium Fungal disease (*Uncinula necator*), commonly known as powdery mildew.

OIV: Organisation Internationale de la Vigne et du Vin An intergovernmental organisation, regarded as the science and technical reference of the vine and wine world. Its member states account for over 85% of world wine production, but excludes the USA, Canada, Mexico and China.

Oloroso A style of Sherry that has been matured in casks that are not completely filled, and so the wine is deliberately oxidised.

Optical sorting table Fast and effective method for sorting grapes that replaces hand sorting tables. The electronic eye analyses the images of the grapes as they move along a belt and compares with previously defined standards to accept or reject.

Oxidation Result of air contact with wine. Oxidised wines will be dried out and bitter.

Pasteurisation Process of heating wine in order to ensure stability by destruction of micro-organisms.

Pectolytic enzymes Proteins used to break down and destroy pectin haze, in order to improve the clarification in wine.

Pedicel The stalk of a single flower or berry.

Peduncle The stem that supports the flower or grape cluster.

Peronospera Fungal disease (*Plasmopara viticola*) affecting vines, commonly known as downy mildew.

Petillance Small amounts of CO_2 present in wine that give a light sparkle.

Petiole Leaf stem.

pH A measure of the hydrogen ion (H^+) concentration in a polar liquid and, hence, the acidity of the liquid. pH is measured on a scale from 0 to 14, where 7 is neutral, below 7 is acidic, and above 7 is alkaline.

Phenol   Basic building block of a group of chemical compounds that include polyphenols q.v.

Phenolic ripeness   Refers to maturity of tannins occurring in grape skins, seeds and stalks that contribute to colour, flavours and aromas. Also known as physiological ripeness.

Phloem   Tubular structures of specialised plant cells that carry sugars derived from photosynthesis and other products of plant leaf metabolism, from the leaves of vines to the trunks and roots and, importantly, to the grapes where sugars are stored.

Photosynthesis   A biosynthetic process of plants by which carbon dioxide and water are combined to form carbohydrates (starches and sugars). Chlorophyll, the compound that pigments plant leaves, is vital for photosynthesis, utilising energy from the sunlight.

Phylloxera vastatrix (*Daktulosphaira vitifoliae*)   Root-burrowing aphid that first devastated Europe's vineyards during the latter part of the nineteenth century.

Physiological ripeness   See **Phenolic ripeness**.

Picogram   One trillionth of a gram.

Pigeage   Punching down of the cap of grape skins formed during fermentation to prevent drying out and encourage the release of colouring matter and tannins. It can be done either manually or mechanically.

Pipe   Cask of 534 or 550-litre capacity, used in the Douro Valley in Portugal for Port production.

Polymerisation   The process of forming together small, tannin molecules into long-chain tannins. This process is encouraged during 'délestage' q.v. and gives softer and more supple tannins.

Polyphenols   Group of chemical compounds present on grape skins, which includes anthocyanins q.v. and tannins q.v.

Press wine   Wine obtained by pressing the grape skins after maceration. This may be used in blending.

Pressing   Process in winemaking in which the juice is pressed from the skins and other solid matter of the grapes.

Protected Designation of Origin (PDO)   The highest classification under European wine laws.

Pumping over   See **Remontage**.

Pupitre   Wooden easels into which bottles are placed (neck first) for the rémuage process q.v. in sparkling winemaking.

Pyrazines   Chemical compounds that give green, peppery aromas and a herbaceous character in some wines. Cabernet Sauvignon and Sauvignon Blanc are two of the grape varieties that can demonstrate these characteristics, both in aroma and on the palate.

Quinta   An inn or country house but often used as a name for a wine estate in Portugal.

Rachis   The main axis stem of a grape cluster, to which pedicels are attached.

Racking   Transferring juice or wine from one vessel to another, leaving behind any lees or sediment and in so doing clarifying the liquid.

Rectified concentrated grape must (RCM)   Grape must that has been processed to remove all non-sugar components. The resultant clear juice can be used for sweetening wine and enriching must.

Reduction/reductivity   A fault in winemaking resulting in aromas of hydrogen sulfide, mercaptans and disulfides.

Refractometer   An optical instrument used in the vineyard for measuring must weight of grapes to indicate grape ripeness.

Remontage   Process in red winemaking during fermentation when the must is drawn from the bottom of the vat and sprayed onto the floating cap of skins. The purpose is to extract colour, tannins and flavours from the skins. Sometimes an aeration is incorporated.

Rémuage   The process in sparkling winemaking by which the sediment in the bottle is drawn to the neck by twisting and shaking.

Reverse osmosis   A liquid purification technology, using a semi-permeable membrane to remove particles from liquid.

Saignée   Lit. bled. A method of producing rosé wines (and concentrating the colour of red wines). Juice that has acquired the desired amount of colour from maceration upon red grape skins is drawn from a vat after a period of 4–20 h, and continues fermentation like a white wine, that is off the skins.

Seasoning   The period of time during the manufacture of oak barrels, when the planks are allowed to rest in the open, in order to draw out harsh tannins and other unwanted products. It also decreases the possibility of barrel leakage.

Solera   The system used for the maturation and production of Sherry. It involves the use of a large number of casks/barrels where young and older Sherries are stored and then blended to ensure consistency of style.

Spätlese   Lit. 'late harvest' – a German quality wine category made from late-harvested grapes, and without chaptalisation q.v.

Stelvin®   Brand name for the best-known screw-cap closure.

Stuck fermentation   Alcoholic fermentation that has prematurely stopped before the conversion of all sugars.

Sucrose   Not a natural constituent of grapes but used in the chaptalisation process in winemaking q.v.

Sugar   Present in grapes as fructose and glucose. Must enrichment may be in the form of sucrose.

Sulfur dioxide (SO₂)   Chemical compound used widely in winemaking as a preservative, antiseptic and antioxidant.

Süssreserve   Sweet, unfermented grape juice that may be added prior to bottling to sweeten wines. Often practised in Germany and England.

Tailles The final quantity of juice pressed from the grapes during sparkling winemaking, generally considered inferior to the juice from first pressings.

Tannins A loose term encompassing polyphenols that bind and precipitate protein. Present in grape skins, stalks and seeds (condensed tannins) and also oak products (hydrolysable tannins). Tannins have an astringent feel in the mouth, particularly on the gums.

Tartrates Crystal deposits of either potassium bitartrate or calcium tartrate that may form during the maturation or storage period. Potassium bitartrate is naturally present in all wine; most is removed before bottling, but some may linger in the form of harmless tartrate crystals.

Teinturier Vines producing grapes with red flesh, sometimes used to add colour to pale wines. Alicante Bouschet and Dunkelfelder are two of the better-known varieties.

Terroir A French term bringing together the notion of soil, microclimate, landscape and environment within a particular vineyard site and the resultant effect on the vines.

Traditional Method The method of producing high-quality sparkling wines and Champagne, utilising a second fermentation in bottle and the eventual removal of yeasty sediment by a process of riddling and disgorging.

Ullage The air space between the wine and the roof of the cask, or in bottle, cork.

Varietal Wine made from a single grape variety.

Vat A large vessel that is used for fermentation, storage or ageing of wine. Historically made from wood, but now usually from other materials including cement, concrete and stainless steel. Some vats are closed, others open topped.

Veraison The beginning of grape ripening, when the skin softens, and the colour starts to change. Following veraison, the growth in the size of the berry is mainly due to the expansion rather than the reproduction of cells. Sugars increase, and acidity levels fall.

Vins clairs Still wines resulting from the first fermentation during the sparkling winemaking process.

Vine balance A state when vine shoot growth provides enough, but not too much, vegetative growth and leaf area to ripen the crop level sufficiently.

Vin doux naturel (**VDN**) Lit. naturally sweet wine. A type of southern French wine that has had its fermentation arrested by the addition of grape spirit thus leaving considerable residual sugar. Predominantly made from red grapes (Grenache Noir), the colour and hue vary according to type of ageing.

Vin Jaune Lit. yellow wine. Wine that has been aged in cask for several years without being topped up, resulting in oxidation. A speciality of the Jura region in France.

Vine stress   A result of the vine not receiving adequate water levels during its growing season. Vines may be deliberately stressed in order to concentrate energy into producing fruit, as opposed to shoot and leaf growth.

Vitis   Grape-bearing genus of the Vitaceae family.

Vitis vinifera   European species of the *Vitis* genus from which nearly all of the world's wine is made. There are at least 10000 known varieties, e.g. Chardonnay, Cabernet Sauvignon.

Volatile acidity   Present in all wine, resulting from the oxidation of alcohol to acetic acid (acid in vinegar). Small amounts enhance aroma; excessive amounts cause 'vinegary' smell and taste.

Xylem   Plant tissue that transports water and nutrients

Yeast autolysis   A complex chemical process that takes place after fermentation when the wine spends time on the lees (dead yeast cells). Particularly valued in Champagne making where it adds 'biscuity' aromas and gives 'creamy' mouthfeel.

Yeasts   Single-cell micro-organisms that produce enzymes that convert sugars into alcohol. They are naturally present on grape skins and in an established winery building, or may be added in cultured form.

Bibliography

Allen, M., Bell, S., Rowe, N. and Wall, G. (eds) (2000) *Proceedings ASVO Seminar, Use of Gases in Winemaking*. Australian Society of Viticulture and Oenology, Adelaide.

Allen, M. and Wall, G. (eds) (2002) *Proceedings ASVO Oenology Seminar, Use of Oak Barrels in Winemaking*. Australian Society of Viticulture and Oenology, Adelaide.

Bakker, J. and Clarke, R.J. (2012) *Wine Flavour Chemistry*, 2nd edn. Wiley-Blackwell Publishing, Oxford.

Bartoshuk, L. (1993) Genetic and pathological taste variation: what can we learn from animal models and human disease? In: *The Molecular Basis of Smell and Taste* (eds, Chadwick, D., March, J. and Goode, J.), pp. 251–267. John Wiley & Sons, Chichester.

Bird, D. (2010) *Understanding Wine Technology*, 3rd edn. DBQA Publishing, Nottingham.

Castrioto-Scanderberg, A., Hagberg, G.E., Cerasa, A., *et al.* (2005) The appreciation of wine by sommeliers: a functional magnetic resonance study of sensory integration. *Neuroimage*, **25**, 570–578.

Chatonnet, P., Dubourdieu, D. and Boidron, J.N. (1995) The influence of *Brettanomyces/Dekkera* sp. yeasts and lactic acid bacteria on the ethylphenol content of red wines. *American Journal of Enology and Viticulture*, **46**, 463–468.

Campbell, C. (2004) *Phylloxera*. Harper Perennial, London.

Considine, J. and Frankish, E. (2014) *A Complete Guide to Quality in Small Scale Wine Making*. Academic Press – Elsevier, Oxford.

Crossen, T. (1997) *Venture into Viticulture*. Country Wide Press, Woodend, Australia.

Elliott, T. (2010) *The Wines of Madeira*. Trevor Elliott Publishing, Gosport.

EU (1999) Council Regulation (EC) No. 1493/1999 on the common organisation of the market in wine. *Official Journal of the European Communities*, **L 179**, 1–101.

EU (2008) Council Regulation (EC) No. 479/2008 on the common organisation of the market in wine, amending Regulations (EC) No. 1493/1999, (EC) No. 1782/2003, (EC) No. 1290/2005, (EC) No. 3/2008 and repealing Regulations (EEC) No. 2392/86 and (EC) No. 1493/1999. *Official Journal of the European Union Communities*, **L 148**, 1–61.

EU (2009a) Council Regulation (EC) No. 491/2009 amending Regulation (EC) No. 1234/2007 establishing a common organisation of agricultural markets and on specific provisions for certain agricultural products (Single CMO Regulation). *Official Journal of the European Communities*, **L 154**, 1–82.

EU (2009b) Commission Regulation (EC) No. 606/2009 laying down certain detailed rules for implementing Council Regulation (EC) No. 479/2008 as regards the categories of grapevine products, oenological practices and the applicable restrictions. *Official Journal of the European Communties*, **L 193**, 1–59.

EU (2011a) Commission Regulation (EC) No. 538/2011 amending Regulation (EC) No. 606/2009 laying down certain detailed rules for implementing Council Regulation

Wine Production and Quality, Second Edition. Keith Grainger and Hazel Tattersall.
© 2016 John Wiley & Sons, Ltd. Published 2016 by John Wiley & Sons, Ltd.

(EC) No. 479/2008 as regards the categories of grapevine products, oenological practices and the applicable restrictions. *Official Journal of the European Communities*, **L 147**, 6–12.

EU (2011b) Commission Regulation (EU) No. 670/2011 amending Regulation (EC) No. 607/2009 laying down certain detailed rules for the implementation of Council Regulation (EC) No. 479/2008 as regards protected designations of origin and geographical indications, traditional terms, labelling and presentation of certain wine sector products. *Official Journal of the European Communities*, **L 183**, 6–13.

EU (2012) Commission Regulation (EU) No. 203/2012 amending Regulation (EC) No. 889/2008 laying down detailed rules for the implementation of Council Regulation (EC) No. 834/2007, as regards detailed rules on organic wine. *Official Journal of the European Communities*, **L 71**, 42–47.

Foulonneau, C. (2002) *Guide Practique de la Vinification*. 2nd edition Dunod, Paris.

Frankel, C. (2014) *Land and Wine: The French Terroir*. The University of Chicago Press, Chicago.

Galet, P. (2000a) *General Viticulture*. Oenoloplurimédia, Chaintré.

Galet, P. (2000b) *Grape Diseases*. Oenoloplurimédia, Chaintré.

Galet, P. (2000c) *Grape Varieties and Rootstock Varieties*. Oenoloplurimédia, Chaintré.

Gladstones, J. (2011) *Wine, Terroir and Climate Change*. Wakefield Press, Adelaide.

Halliday, J. and Johnson, H. (2006) *The Art and Science of Wine*, 2nd edn. Mitchell Beazley, London.

Hanson, A. (1982) *Burgundy*. Faber, London.

Iland, P., Bruer, N., Edwards, G., Weeks, S. and Wilkes, E. (2004a) *Chemical Analysis of Grapes and Wine: Techniques and Concepts*. Winetitles, Adelaide.

Iland, P., Bruer, N., Ewart, A., Markides, A. and Sitters, S. (2004b) *Monitoring the Winemaking Process from Grapes to Wine: Techniques and Concepts*. Winetitles, Adelaide.

Iland, P., Grbin, P., Grinbergs, M., Schmidtke, L. and Soden, A. (2007) *Microbiological Analysis of Grapes and Wine: Techniques and Concepts*. Winetitles, Adelaide.

ITV (1991) *Protection Raisonnée du Vignoble*. Centre Technique Interprofessionnel de la Vigne et du Vin, Paris.

ITV (1995) *Guide d'Établissement du Vignoble*. Centre Technique Interprofessionnel de la Vigne et du Vin, Paris.

ITV (1998) *Matérials et Installations Vinicoles*. Centre Technique Interprofessionnel de la Vigne et du Vin, Paris.

Jackson, R.S. (2008) *Wine Science (Principles and Applications)*, 3rd edn. Academic Press – Elsevier, San Diego.

Johnson, J. and Robinson, J. (2013) *The World Atlas of Wine*, 7th edn. Octopus Publishing, London.

Karlsson, B. and Karlsson, P. (2014) *Biodynamic, Organic and Natural Winemaking*. Floris Books, Edinburgh.

Legeron, I. (2014) *Natural Wine: An Introduction to Organic and Biodynamic Wines Made Naturally*. Cico Books – Ryland Peters & Small, London.

Magarey, P., MacGregor, A.M., Wachtel, M.F. and Kelly, M.C. (2000 – reprinted 2013) *The Australian and New Zealand Field Guide to Diseases, Pests and Disorders of Grapes*. Winetitles, Adelaide.

Margalit, Y. (2013) *Concepts in Wine Technology*, 3rd edn. The Wine Appreciation Guild, San Francisco.

Michelsen, C.S. (2005) *Tasting & Grading Wine*. JAC International, Limhamn.

Ministry of Agriculture, Fisheries and Food (1997) *Catalogue of Selected Wine Grape Varieties and Clones Cultivated in France*. ENTAV (Établissement National Technique pour l'Amélioration de la Viticulture), INRA (Institut National de Recherche Agronomique), ENSAM (École Nationale Supérieure Agronomique de Montpellier), ONIVINS (Office National Interprofessionnel des Vins).

Nicholas, P. (ed.) (2004) *Soil, Irrigation and Nutrition*. South Australian Research and Development Institute, Adelaide.

Nicholas, P., Magarey, P. and Wachtel, M. (eds) (1994) *Diseases and Pests, Grape Production Series Number 1*. Winetitles, Adelaide.

OIV (2015a) *International Code of Oenological Practices*. OIV – Organisation Internationale de la Vigne et du Vin, OIV, Paris.

OIV (2015b) *International Oenological Codex*. OIV – Organisation Internationale de la Vigne et du Vin, OIV, Paris.

Parker, R. (2005) *The World's Greatest Wine Estates*. Dorling Kindersley, London.

Penning Rowsell, E. (1979) *The Wines of Bordeaux*. 4th edn. Penguin, Harmondsworth.

Peynaud, E. (1987) *The Taste of Wine*. John Wiley, New York.

Rankine, B. (2004) *Making Good Wine*. Macmillan, Sydney.

Redding, C. (1833) *The History and Description of Modern Wines*. Whittaker, Treacher, & Arnot, London

Reynolds, A.G. (ed.) (2010a) *Managing Wine Quality Vol. 1. Viticulture and Wine Quality*. Woodhead Publishing, Cambridge.

Reynolds, A.G. (ed.) (2010b) *Managing Wine Quality Vol. 2. Oenology and Wine Quality*. Woodhead Publishing, Cambridge.

Ribéreau-Gayon, P., Glories, Y, Maujean, A. and Dubourdieu, D. (2006a) *Handbook of Enology Vol. 2 – The Chemistry of Wine: Stabilization and Treatments*. John Wiley & Sons, Chichester.

Ribéreau-Gayon, P., Dubourdieu, D., Donèche, B. and Lonvaud, A. (2006b) *Handbook of Enology Vol. 1 – The Microbiology of Wine and Vinifications*. John Wiley & Sons, Chichester.

Robinson, J. (ed.) (2015) *The Oxford Companion to Wine*, 4th edn. Oxford University Press, Oxford.

Robinson, J., Harding, J. and Vouillamoz, J. (2012) *Wine Grapes*. Allen Lane, London.

Saintsbury, G. (1920) *Notes from a Cellar Book*. Macmillan, London.

Schuster, M. (1989) *Understanding Wine*. Mitchell Beazley, London.

Seguin, G. (1986) 'Terroirs' and pedology of wine growing. *Experentia*, **42**, 861–871.

Smart, R. and Robinson, M. (1991) *Sunlight into Wine*. Winetitles, Adelaide.

Tamine, A.Y. (ed.) (2013) *Membrane Processing – Dairy and Beverage Applications*. Wiley-Blackwell, Chichester.

Vigne et Vin (2003) *Guide Pratique, Viticulture Biologique*. Vigne et Vin, Bordeaux.

Vigne et Vin (2004) *The Barrel: Selection, Utilization, Maintenance*. Vigne et Vin, Bordeaux.

Wilson, J. (1998) *Terroir*. Mitchell Beazley, London.

Wine Australia (2015) *Wine Exports Approval Report, December 2014*. Wine Australia, Adelaide.

Useful websites

This list contains a selection of websites that the authors consider to be well constructed and contain valuable information for the reader. It is, by definition, far from exhaustive.

American Association of Wine Economists (AAWE): http://www.wine-economics.org
The AAWE is a not-for-profit organisation based in New York. It publishes a journal three times a year, and the website includes links to working papers broadly related to topics of wine economics.

Association of Wine Educators: http://www.wineeducators.com
The Association of Wine Educators is a professional association whose members are involved in the field of wine education. The website includes a directory of members, many of whom are specialists in different aspects of wine production and assessment.

The Australian Wine Research Institute: http://www.awri.com.au
The Australian Wine Research Institute, whose aim is to advance the competitive edge of the Australian Wine Industry, conducts research into the composition and sensory characteristic of wines, undertakes an analytical service and provides industry development and support. The website contains a good deal of free information.

Bordeaux Wines: CIVB Conseil Interprofessionnel du Vin de Bordeaux: http://www.bordeaux.com/uk
The CIVB represents, advises and controls the wine industry of Bordeaux, which is the largest fine-wine region in the world. This is a lively website with much useful information about Bordeaux and its wines.

Circle of Wine Writers: http://www.circleofwinewriters.org
The Circle of Wine Writers is an association of wine writers and communicators with members all around the world.

Decanter: http://www.decanter.com
Decanter is a monthly wine magazine, aimed at consumers but widely read by the wine trade and wine producers. The website contains many topical news items.

The Drinks Business: http://www.thedrinksbusiness.com
The Drinks Business is a monthly magazine that focuses upon business aspects of the alcoholic drinks industry. There are comprehensive reports on wine regions, trends, brands and in-depth discussions of challenges.

Wine Production and Quality, Second Edition. Keith Grainger and Hazel Tattersall.
© 2016 John Wiley & Sons, Ltd. Published 2016 by John Wiley & Sons, Ltd.

Grainger, Keith: http://www.keithgrainger.com
This website contains contact information about the lead author of this book.

Harpers: http://www.harpers.co.uk
Harpers is a journal for the UK wine trade. The website includes a précis of some recent articles.

The Institute of Masters of Wine: http://www.masters-of-wine.org
The qualification of Master of Wine (MW) is regarded as the highest level of achievement in broadly based wine education. The website includes a list of the Institute's members.

Institut Français de la Vigne et du Vin: http://www.vignevin.com/
The institute provides technical information for growers and winemakers in France. The website (in French) includes details of their publications.

Organisation Internationale de la Vigne et du Vin (OIV): http://www.oiv.int
The OIV is an intergovernmental organisation and is regarded as the science and technical reference of the vine and wine world. Its member states account for over 85% of world wine production, but excludes the USA, Canada, Mexico and China. The website provides up-to-date news and many useful statistics.

New Zealand Winegrowers: http://www.nzwine.com
The New Zealand Winegrowers website contains valuable information about the industry, including statistics, profiles, reports and links to producers.

Plumpton College: http://www.plumpton.ac.uk/department/wine-and-wine-research/21
Plumpton College based in Sussex, UK offers many wine courses including full and part-time degrees in viticulture, oenology and wine business.

TiZwine: http://www.tizwine.com
Based in New Zealand, TiZwine is a valuable source of information on the country's wines and news on the world of wine generally.

UC Davis: http://wineserver.ucdavis.edu
The Department of Viticulture and Enology at Davis has been at the forefront of research for 40 years.

Université Victor Segalen Bordeaux 2, Faculté d'Oenologie: http://www.isvv.univ-bordeauxsegalen.fr/en
Institute of Vine and Wine Science and a centre of excellence for studies of oenology in France.

Wiley: http://www.wiley.com
The website of this book's publishers, which includes details of publications and links to numerous resources.

Wines of Chile: http://www.winesofchile.org
Wines of Chile represents 90 Chilean wineries. The website is a useful source of news, statistics and reports.

Wines of South Africa: http://www.wosa.co.za
Wines of South Africa represents exporters of South African wines. The website contains up-to-date news and links to useful statistics.

Wine & Spirit Education Trust: http://www.wsetglobal.com
The Wine & Spirit Education Trust designs and provides wine courses, and is an industry awards body whose qualifications are recognised by the Qualifications and Curriculum Authority in the United Kingdom under the Qualifications and Credit Framework. The Trust is international, and there are Approved Programme Providers in over 60 countries worldwide.

Winetitles: http://www.winetitles.com.au
Winetitles is an Australian publisher of wine books, journals and seminar papers.

Wine and vineyard & winery equipment exhibitions

Listed below are some of the most important wine and viticulture/winemaking equipment exhibitions in Europe and Australasia. For those wishing to widen their tasting experience, the wine shows represent a valuable opportunity to taste a huge array of wines under one roof in a short space of time. For producers and students of viticulture or oenology, the equipment exhibitions are an invaluable way of keeping abreast of new developments.

United Kingdom

London Wine Fair
Held in London at the Olympia Exhibition Centre on Tuesday, Wednesday and Thursday, normally in the third week of May. The fair is a trade-only exhibition. http://www.londonwinefair.com

France

Vinexpo
Held in Bordeaux biannually in June, this is a huge, trade-only, wine exhibition. http://www.vinexpo.fr

Vinitech-Sifel
Held in Bordeaux biannually in late November/early December of the years when Vinexpo does not take place, this is the major viticulture and winemaking exhibition in France. http://www.vinitech-sifel.com/

Germany

Prowein
Held in Düsseldorf over four days in mid-March, this trade-only exhibition has gained in importance in recent years and is considered by many to be the No. 1 wine-trade exhibition in Europe. http://www.prowein.com

Wine Production and Quality, Second Edition. Keith Grainger and Hazel Tattersall.
© 2016 John Wiley & Sons, Ltd. Published 2016 by John Wiley & Sons, Ltd.

Italy

Enovitis
Held in Milan in November, this is the most important vineyard and winery equipment exhibition in Italy. http://enovitis.net

Vinitaly
Held in Verona over four days in early April, this trade-only exhibition is by far the most important Italian wine event. http://www.vinitaly.com

Spain

Fenavin
Held in Ciudad Real in May, this recently established exhibition focuses on the wines of Spain. http://www.fenavin.com

China

Vinexpo Asia Pacific
Held in Hong Kong over four days in late May, this exhibition is the major trade wine show in the Far East. http://www.vinexpo.com

Australia

Winetech
Held in Adelaide, this is the major viticulture and winemaking exhibition in Australia. http://www.winetechaustralia.com.au

Index

abscisic acid, 247
AC *see* Appellation Contrôlée
Accad, Guy, 98
acetaldehyde, 102
  excessive, 209
acetic acid, 10, 75, 182, 184, 205,
    208, 269
acetic spoilage, 76
acetobacter, 205
acid, 9
  abscisic, 247
  acetic, 10, 75, 182, 184, 205, 208, 269
  ascorbic, 184, 269
  citric, 10, 184
  gluconic, 184
  lactic, 75, 90, 184, 203, 205
  malic, 9, 50, 84–85, 90, 97, 184,
    205, 269
  metatartaric, 122
  sorbic, 184, 269
  succinic, 10, 75, 184
  tartaric, 9, 50, 84–85, 121, 165, 174,
    184, 205
acidex®, 85
acidification, 24, 85, 279
acidity, 9, 69, 84–85, 148, 167, 189
  impact of climate, 17, 24, 70, 225–226
  tasting, 154, 158, 162, 182–185, 189
  total, 84, 148, 205, 269–270
  volatile, 102, 104, 121, 201, 205, 269
Aconcagua, 38
aeration, 88–89, 98–99, 102, 160
aftertaste, 150, 181, 191
ageing potential, 94, 159, 165, 197
agrafe, 139
albariza, 32, 131
albumin, 117

alcohol, 9, 17, 75–76, 86, 131–134,
    268–269
  affect of climate, 17, 21, 23–24,
    186, 248
  ethyl, 3, 75
  evaporation, 114–115
  reduction, 98, 103, 105
  tasting, 162, 172–173, 182–183,
    186–187
aldehyde, 177, 180
Alentejo, 46
Alicante Bouschet, 167
Alsace, 6, 24, 64, 137, 144
alte reben, 244
Amarone della Valpolicella, 128–129, 165
Amerine and Winkler, 23
amino acids, 10, 178, 182
Amontillado, 132
Ampelidaceae, 5
amphorae, 78, 109–110, 254
animals
  (pests), 53
  (working), 37, 63
annual growth cycle, 24, 47–49
anosmics, 149
anthocyanins, 8, 50, 70, 100–101
anthracnose, 56
AOP *see* Appellation d'Origine
    Protégée
appassimento, 128–129
Appellation Contrôlée, 45–46, 105, 137,
    212, 214
Appellation d'Origine Protégée, 45–46,
    105, 212
argilo calcaire, 227
arginine, 10
argon, 256

Wine Production and Quality, Second Edition. Keith Grainger and Hazel Tattersall.
© 2016 John Wiley & Sons, Ltd. Published 2016 by John Wiley & Sons, Ltd.

Printed and bound by CPI Group (UK) Ltd, Croydon, CR0 4YY

27/10/2024

14580351-0001